Master HiSET®

Math

Math Practice, Tutorials and Multiple Choice Strategies

AF722177

COMPLETE
TEST PREPARATION INC.
WWW.TEST-PREPARATION.CA

Copyright © 2025 by Complete Test Preparation Inc. ALL RIGHTS RESERVED.

No part of this book may be reproduced or transferred in any form or by any means, graphic, electronic, or mechanical, including photocopying, recording, web distribution, taping, or by any information storage retrieval system, without the written permission of the author.

Notice: Complete Test Preparation Inc. makes every reasonable effort to obtain from reliable sources accurate, complete, and timely information about the tests covered in this book. Nevertheless, changes can be made in the tests or the administration of the tests at any time and Complete Test Preparation Inc. makes no representation or warranty, either expressed or implied as to the accuracy, timeliness, or completeness of the information contained in this book. Complete Test Preparation Inc. makes no representations or warranties of any kind, express or implied, about the completeness, accuracy, reliability, suitability or availability with respect to the information contained in this document for any purpose. Any reliance you place on such information is therefore strictly at your own risk.

The author(s) shall not be liable for any loss incurred as a consequence of the use and application, directly or indirectly, of any information presented in this work. Sold with the understanding, the author is not engaged in rendering professional services or advice. If advice or expert assistance is required, the services of a competent professional should be sought.

HiSET® is a registered trademark of the Educational Testing Service (ETS), who are not involved in the production of, and do not endorse this product.

We strongly recommend that students check with exam providers for up-to-date information regarding test content.

Published by
Complete Test Preparation Inc.
Victoria BC Canada

Visit us on the web at
https://www.test-preparation.ca
Printed in the USA

ISBN-13:

Version 8.5 Updated August 2025

About Complete Test Preparation Inc.

Why Us?
The Complete Test Preparation Team has been publishing high quality study materials since 2005, with a catalogue of over 145 titles, in English, French, Spanish and Chinese, as well as ESL curriculum for all levels.

To keep up with the industry changes, we update everything all the time!

And the best part?
With every purchase, you're helping people all over the world improve themselves and their education. So thank you in advance for supporting this mission with us! Together, we are truly making a difference in the lives of those often forgotten by the system.

Charities that we support
https://www.test-preparation.ca/charities-and-non-profits/

You have definitely come to the right place.
If you want to spend your valuable study time where it will help you the most - we've got you covered today and tomorrow.

Feedback

We welcome your feedback. Email us at feedback@test-preparation.ca with your comments and suggestions. We carefully review all suggestions and often incorporate reader suggestions into upcoming versions. As a Print on Demand Publisher, we update our products frequently.

Contents

8 Getting Started

The HiSET© Math Content	9
The HiSET© Study Plan	10
Making a Study Schedule	10
Tips for making a schedule	16

18 Basic Math

Basic math Video Tutorials	18
Fraction Tips, Tricks and Short-cuts	18
Decimal Tips, Tricks and Short-cuts	25
Converting Decimals to Fractions	25
Percent Tips, Tricks and Short-cuts	27
Most Common Basic Math Mistakes	29
Scientific Notation	30
How to Convert to Scientific Notation	30
Exponents and Radicals	32
Exponents Tips, Short-cuts & Tricks	32
Simplifying Radicals	34
Basic Math Practice Questions	37
Answer Key	50

59 Data, Statistics and Probability

Mode, Mean and Median	59
Dependent and Independent Variables	60
Concepts in Statistics and Probability	61
Data Analysis and Statistics Practice	64
Answer Key	72

75 Word Problems

How to Solve Word Problems	75
Types of Word Problems	78
Word Problem Practice with Video	90
Most Common Word Problem Mistakes	90
Word Problem Practice	94
Answer Key	102

113 **Basic Geometry**
Pythagorean Geometry 121
Quadrilaterals 123
Congruence 126
Fundamental Theorems in Geometry 130
Important Equations in Geometry 131
Geometry Practice Questions 133
Answer Key 156

170 **Algebra**
Solving One-Variable Linear Equations 170
Solving Two-Variable Linear Equations 171
Algebra Solutions - Video Tutorials 173
Common Mistakes 176
Simplifying Polynomials 178
Factoring Polynomials 178
Quadratic Equations 180
Quadratic Word Problems 181
Basic Algebra Practice 185
Answer Key 198

218 **Basic Math Multiple Choice**
Math Multiple Choice Strategy 227

224 **How to Study for a Math Test**
Quick Tips - Studying for a Math Test 233

236 **How to Prepare for a Test**
Test Prep - the Basics 237
The Strategy of Studying 238

242 **How to Take a Test**
Reading the Instructions 242
How to Take a Test - The Basics 244
In the Test Room – What you MUST do! 248
Avoid Anxiety Before a Test 254
Common Test-Taking Mistakes 256

259 **Conclusion**

260 **Online Resources**

Getting Started

CONGRATULATIONS! By deciding to take the HiSET© you have taken the first step toward a great future! Of course, there is no point in taking this important examination unless you intend to do your best to earn the highest grade that you possibly can. That means getting yourself organized and discovering the best approaches, methods and strategies to master the material. Yes, that will require real effort and dedication on your part, but if you are willing to focus your energy and devote the study time necessary, before you know it you will be passing the HiSET© with a great score!

We know that taking on a new endeavor can be scary, and it is easy to feel unsure of where to begin. That's where we come in. This workbook is designed to help you improve your test-taking skills, show you a few tricks of the trade and increase both your competency and confidence.

The HiSET© Math Content

Basic Math

Exponents and Radicals Square Root
Fractions, Decimals and Percent
Order of Operations
Word Problems
Metric Conversion

Algebra

Ratio and proportion
Linear equations with 1 and 2 variables
Quadratics
Real-world quadratic problems
Identify quadratic equations from graphs
Solve Inequalities
Monomials
Algebraic geometry problems
Basic Trigonometry

Simple Geometry

Area, Volume and Perimeter
Slope of a line
Identify linear equations from a graph
Calculate perimeter, circumference and volume
Solve problems using the Pythagorean theorem
Solve real world problems using the properties of geometric shapes
Similarity and Congruence

Data Analysis and Statistics

Means, Median and Mode
Dependent and Independent Variables Interpreting Graphs and Tables

The HiSET© Study Plan

Now that you have made the decision to take the HiSET©, it is time to get started. Before you do another thing, you will need to figure out a plan of attack. The very best study tip is to start early! The longer the time period you devote to regular study practice, the more likely you will be to retain the material and be able to access it quickly. If you thought that 1 x 20 is the same as 2 x 10, guess what? It really is not, when it comes to study time. Reviewing material for just an hour per day over the course of 20 days is far better than studying for two hours a day for only 10 days. The more often you revisit a particular piece of information, the better you will know it. Not only will your grasp and understanding improve, but your ability to reach into your brain and quickly and efficiently pull out the tidbit you need, will be greatly enhanced as well.

The great Chinese scholar and philosopher Confucius believed that true knowledge could be defined as knowing both what you know and what you do not know. The first step in preparing for the HiSET© is to assess your strengths and weaknesses. You may already have an idea of what you know and what you do not know, but evaluating yourself for each of the math areas will clarify the details.

Making a Study Schedule

To make your study time the most productive, you will need to develop a study plan. The purpose of the plan is to organize all the bits of pieces of information in such a way that you will not feel overwhelmed. Rome was not built in a day, and learning everything you will need to know to pass the HiSET© is going to take time, too. Arranging the material you need to learn into manageable chunks is the best way to go. Each study session should make you feel as though you have succeeded in accomplishing your goal, and your goal is simply to learn what you planned to learn during that particular session. Try to organize the content in such a way that each study session builds on previous ones. That way, you will retain the information, be better able to access it, and review the previous bits and pieces at the same time.

Self-assessment

The Best Study Tip! The very best study tip is to start early! The longer you study regularly, the more you will retain and 'learn' the material. Studying for 1 hour per day for 20 days is far better than studying for 2 hours for 10 days.

What don't you know?

The first step is to assess your strengths and weaknesses. You may already have an idea of where your weaknesses are, or you can take our Self-assessment modules for each of the areas, Reading Comprehension, Arithmetic, Essay Writing, Algebra and College Level Math.

Subject	Rate 1 - 5
Basic Math	
Exponents, Radicals and Square Root	
Fractions, Decimals and Percent	
Order of Operations	
Word Problems	
Metric Conversion	

Algebra	
Ratio and proportion	
Linear equations with 1 and 2 variables	
Quadratics	
Real-world quadratic problems	
Identify quadratic equations from graphs	
Solve Inequalities	
Monomials	
Algebraic geometry problems	
Basic Trigonometry	
Simple Geometry	
Area, Volume and Perimeter	
Slope of a line	
Identify linear equations from a graph	
Calculate perimeter, circumference and volume	
Solve problems using the Pythagorean theorem	
Similarity and Congruence	
Data Analysis and Statistics	
Means, Median and Mode	
Interpreting Graphs and Tables	

Dependent and Independent Variables	

Making a Study Schedule

The key to a study plan is to divide the material you need to learn into manageable size and learn it, while at the same time reviewing the material that you already know.

Using the table above, any scores of three or below, you need to spend time learning, going over, and practicing this subject area. A score of four means you need to review the material, but you don't have to spend time re-learning. A score of five and you are OK with just an occasional review before the exam.

A score of zero or one means you really do need to work on this and you should allocate the most time and give it the highest priority. Some students prefer a 5-day plan and others a 10-day plan. It also depends on how much time until the exam.

Here is an example of a 5-day plan based on an example from the table above:

Trigonometry: 1 Study 1 hour everyday – review on last day
Fractions: 3 Study 1 hour for 2 days then ½ hour and
then review
Algebra: 4 Review every second day
Monomials: 2 Study 1 hour on the first day – then ½ hour everyday
Geometry: 5 Review for ½ hour every other day

Using this example, geometry and sequences are good and only need occasional review. Algebra is good and needs 'some' review. Fractions need a bit of work, Monomials needs a lot of work and trigonometry is very weak and need most of time. Based on this, here is a sample study plan:

Day	Subject	Time
Monday		
Study	Trigonometry	1 hour
Study	Monomials	1 hour
	½ hour break	
Study	Fractions	1 hour
Review	Algebra	½ hour
Tuesday		
Study	Trigonometry	1 hour
Study	Monomials	½ hour
	½ hour break	
Study	Fractions	½ hour
Review	Algebra	½ hour
Review	Geometry	½ hour
Wednes-day		
Study	Trigonometry	1 hour
Study	Monomials	½ hour
	½ hour break	
Study	Fractions	½ hour
Review	Geometry	½ hour
Thursday		
Study	Trigonometry	½ hour
Study	Monomials	½ hour
Review	Fractions	½ hour
	½ hour break	
Review	Geometry	½ hour
Review	Algebra	½ hour
Friday		
Review	Trigonometry	½ hour
Review	Monomials	½ hour
Review	Fractions	½ hour
	½ hour break	
Review	Algebra	½ hour
Review	Monomials	½ hour

Using this example, adapt the study plan to your own schedule. This schedule assumes 2 ½ - 3 hours available to study everyday for a 5 day period.

First, write out what you need to study and how much. Next figure out how many days before the test.

Note, do NOT study on the last day before the test. On the last day before the test, you won't learn anything and will probably only confuse yourself.

Make a table with the days before the test and the number of hours you have available to study each day. We suggest working with 1 hour and ½ hour time slots.

Start filling in the blanks, with the subjects you need to study the most getting the most time and the most regular time slots (i.e. everyday) and the subjects that you know getting the least time (e.g. ½ hour every other day, or every 3rd day).

Tips for making a schedule

Once you make a schedule, stick with it! Make your study sessions reasonable. If you make a study schedule and don't stick with it, you set yourself up for failure. Instead, schedule study sessions that are a bit shorter and set yourself up for success! Make sure your study sessions are do-able. Studying is hard work, but after you pass, you can party and take a break!

Schedule breaks. Breaks are just as important as study time. Work out a rotation of studying and breaks that works for you.

Build up study time. If you find it hard to sit still and study for 1 hour straight through, build up to it. Start with 20 minutes, and then take a break. Once you get used to 20-minute study sessions, increase the time to 30 minutes. Gradually work you way up to 1 hour.

40 minutes to 1 hour is optimal. Studying for longer than this is tiring and not productive. Studying for shorter isn't long enough to be productive.

More on how to study math
https://www.test-preparation.ca/study-math/

How to Study
For more information, see our How to Study Guide at https://www.test-preparation.ca/learning-study/

Flash Cards - The Complete Guide
https://www.test-preparation.ca/flash-cards/

Using your Daily Routine to Study
https://www.test-preparation.ca/daily-routine/

Basic Math

THE BASIC MATH SECTION COVERS:

- Fractions, Decimals and Percent
- Scientific Notation
- Exponents and Radicals

Basic math Video Tutorials

https://www.test-preparation.ca/basic-math-video-tutorials/
https://test-preparation.ca/basic-math-practice-test-questions/

Fraction Tips, Tricks and Short-cuts

When you are writing an exam, time is precious, so anything you can do to answer questions faster is a real advantage.

Here are some ideas, Short-cuts, tips and tricks that can speed up answering fraction problems.

Remember that a fraction is just a number which names a portion of something. For instance, instead of having a whole pie, a fraction says you have a part of a pie--such as a half of one or a fourth of one.

Two numbers make up a fraction. The number on top is the numerator. The number on the bottom is the denominator.

To remember which is which, just remember that "denominator" and "down" both start with a "d." And the "downstairs" number is the denominator. So for instance, in ½, the numerator is 1, and the denominator (or "downstairs") number is 2.

Adding Fractions

It's easy to add two fractions if they have the same denominator. Just add the digits on top and leave the bottom one the same: 1/10 + 6/10 = 7/10.

It's the same with subtracting fractions with the same denominator: 7/10 - 6/10 = 1/10.

Adding and subtracting fractions with different denominators is a little more complicated.

First, you have to arrange the fractions so they have the same denominators.

The easiest way to do this is to multiply the denominators: For 2/5 + 1/2 multiply 5 by 2. Now you have a denominator of 10.

But now you have to change the top numbers too. Since you multiplied the 5 in 2/5 by 2, you also multiply the 2 by 2, to get 4. So the first fraction is now 4/10.

In the second fraction, you multiplied the denominator by 5, you have to multiply the numerator by 5 also, to get 5/10.

Now you have 4/10 + 5/10 and you can add 5 and 4 to get 9/10.

Simplest Form

To reduce a fraction to its simplest form, you have to arrange the numerator and denominator so the only common factor is 1.

Think of it this way:

Let's take an example: The fraction 2/10.

This is not reduced to its simplest terms because there is a number that will divide evenly into both: 2. We want to make it so that the only number that will divide evenly into both is 1.

Divide the top and bottom by 2 to get the new, reduced fraction - 1/5.

Multiplying Fractions

This is the easiest of all: Just multiply the two top numbers and then multiply the two bottom numbers.

Here is an example,

2/5 X 2/3

First, multiply the numerators: 2 X 2 = 4

then multiply the denominators: 5 X 3 = 15

Your answer is 4/15.

Dividing Fractions

Dividing fractions is easy if you remember a simple trick - first turn the second fraction upside down - then multiply!

Here is an example:

7/8 X 1/2

Turn the second fraction upside down:

7/8 X 2/1

then multiply:
(7 X 2) / (8 X 1) = 14/8

Converting Fractions to Decimals

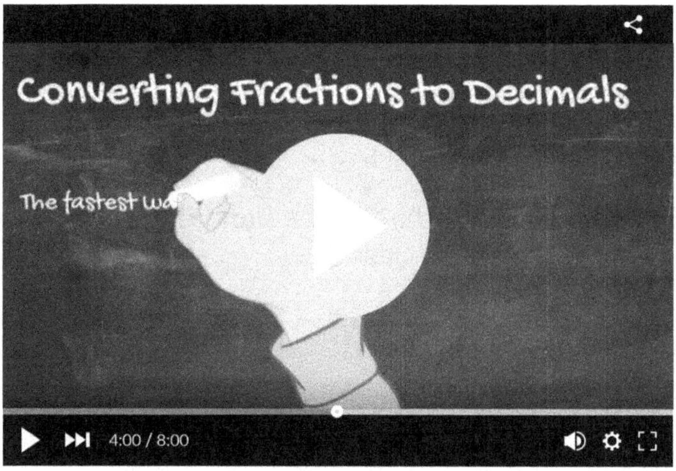

https://youtu.be/0OA9yXAc4fY

There are a couple of ways to convert fractions to decimals. The first, which is the fastest -- is to memorize some basic fraction facts.

1/100 is "one hundredth," expressed as a decimal, it's .01.

1/50 is "two hundredths," expressed as a decimal, it's .02.

1/25 is "one twenty-fifth" or "four hundredths," expressed as a decimal, it's .04.

1/20 is "one twentieth" or """five hundredths," expressed as a decimal, it's .05.

1/10 is "one tenth," expressed as a decimal, it's .1.

1/8 is "one eighth," or "one hundred twenty-five thousandths," expressed as a decimal, it's .125.

1/5 is "one fifth," or "two tenths," expressed as a decimal, it's .2.

1/4 is "one fourth" or "twenty-five hundredths," expressed as a decimal, it's .25.

1/3 is "one third" or "thirty-three hundredths," expressed as a decimal, it's .33.

1/2 is "one half" or "five tenths," expressed as a decimal, it's .5.

3/4 is "three fourths," or "seventy-five hundredths," expressed as a decimal, it's .75.

Of course, if you're no good at memorization, another good technique for converting a fraction to a decimal is to manipulate it so that the fraction's denominator is 10, 100, 1000, or some other power of 10.

Here's an example: We'll start with three quarters. What is the first number in the 4 "times table" that you can multiply and get a multiple of 10? Can you multiply

4 by something to get 10? No. Can you multiply it by something to get 100? Yes! 4 X 25 is 100.

So multiply the numerator by 25, which is 75 over 100

We know fractions are really a division problem, and we also know that dividing by 100, means we move the decimal 2 places to the left.

So, 75 over 100 = .75

Lets try another example - Convert one fifth to a decimal.

First find a power of 10 that 5 goes into evenly, which is 2.

Multiply the numerator and denominator by 2, which is

two tenths.

Dividing 2 by 10 means we move the decimal place 1 place to the left.

So 1/5 = 0.5

Converting Fractions to Percent

Here is a quick method to convert fraction to percent and a strategy for answering on a multiple choice test that will save you valuable exam time.

First, remember that a fraction is a division problem: you're dividing the bottom number into the top.

Taking an example, convert 2/3 into percent.

The first method is to multiple the numerator by 100 and divide. So,

(2 X 100) / 2 = 100/3 = 66.66

Add a % sign and you have the answer, 66.66%

If you're doing these conversions on a multiple-choice test, here's an idea that might be even easier and faster. Let's say you have a fraction of 1/8 and you're asked to convert to percent.

Since we know that "percent" means hundredths, ask yourself what number we can multiply 8 by to get 100. Since there is no number, ask what number gets us close to 100.

That number is 12: 8 X 12 = 96. So it gets us a little less than 100. Now, whatever you do to the denominator, you have to do to the numerator. Let's multiply 1 X 12 and we get 12. However, since 96 is a little less than 100, we know that our answer will be a little MORE than 12%.

Look at the choices and eliminate the obvious wrong choices. So if your possible answers on the multiple-choice test are these:

a) 8.5% b) 19% c)12.5% d) 25%

then we know the answer is c) 12.5%, because it's a little MORE than the 12 we got in our math problem above. Here all the choices except choice C 12.5% can be eliminated.

You don't have to know the exact correct answer, just enough to estimate, then eliminate the obviously wrong answers.

This was an easy example to demonstrate the strategy, but don't be fooled! You probably won't get such an easy question on your exam. By estimating your answer quickly, then eliminating obviously incorrect choices immediately, you save precious exam time.

Decimal Tips, Tricks and Short-cuts

HTTPS://YOUTU.BE/4KTB4CK1SO8

Converting Decimals to Fractions

Converting decimals to fractions is easy if you say it the right way! If you say "point one" or "point 25," you'll have trouble.

But if you say, “one tenth” and “twenty-five hundredths,” then you have already solved it! That’s because, if you know your fractions, you know that “one tenth” looks like this: 1/10. And “twenty-five hundredths” looks like this: 25/100.

Even if you have digits before the decimal, such as 3.4, learning how to say the word will help you with the conversion into a fraction. It’s not “three point four,” it’s “three and four tenths.” Knowing this, you know that the fraction which looks like “three and four tenths” is 3 4/10.

The conversion is not complete until you reduce the fraction to its lowest terms: It’s not 25/100, but 1/4.

Converting Decimals to Percent

Changing a decimal to a percent is easy if you remember one thing: multiply by 100.

For example, if you start with .45, simply multiply it by 100 for 45. Then add the % sign to the end - 45%.

Think of it this way: take out the decimal point, add a percent sign on the opposite side. In other words, the decimal on the left is replaced by the % on the right.

It doesn’t work quite that easily if the decimal is in the middle of the number. For example, 3.7. Here, take out the decimal in the middle and replace it with a 0 % at the end. So 3.7 converted to decimal is 370%.

Percent Tips, Tricks and Short-cuts

Percent problems are not nearly as scary as they appear, if you remember this neat trick:

Draw a cross as in:

Portion	Percent
Whole	100

In the upper left, write PORTION. In the bottom left write WHOLE. In the top right, write PERCENT and in the bottom right, write 100. Whatever your problem is, you will leave blank the unknown, and fill in the other four parts. For example, let's suppose your problem is: Find 10% of 50. Since we know the 10% part, we put 10 in the percent corner. Since the whole number in our problem is 50, we put that in the corner marked whole. You always put 100 underneath the percent, so we leave it as is, which leaves only the top left corner blank. This is where we'll put our answer. Now simply multiply the two corner numbers that are NOT 100. Here, it's 10 X 50. That gives us 500. Now divide this by the remaining corner, or 100, to get a final answer of 5. 5 is the number that goes in the upper-left corner, and is your final solution.

Another hint to remember: Percents are the same thing as hundredths in decimals. So .45 is the same as 45 hundredths or 45 percent.

Converting Percents to Decimals

Percents are just a type of decimal, so it should be no surprise that converting between the two is actually fairly simple. Here are a few tricks and Short-cuts to keep in mind:

- Remember that percent literally means "per 100" or "for every 100." So when you speak of 30% you're saying 30 for every 100 or the fraction 30/100. In basic math, you learned that fractions that have 10 or 100 as the denominator can easily be turned to a decimal. 30/100 is thirty hundredths, or expressed as a decimal, .30.
- Another way to look at it: To convert a percent to a decimal, simply divide the number by 100. So for instance, if the percent is 47%, divide 47 by 100. The result will be .47. Get rid of the % mark and you're done.
- Remember that the easiest way of dividing by 100 is by moving your decimal two spots to the left.

Converting Percents to Fractions

Converting percents to fractions is easy. After all, a percent is just a type of fraction; it tells you what part of 100 that you're talking about. Here are some simple ideas for making the conversion from a percent to a fraction:

- If the percent is a whole number -- say 34% -- then simply write a fraction with 100 as the denominator (the bottom number). Then put the percentage itself on top. So 34% becomes 34/100.
- Now reduce as you would reduce any percent. In this case, by dividing 2 into 34 and 2 into 100, you get 17/50.
- If your percent is not a whole number -- say 3.4% --then convert it to a decimal expressed as hundredths. 3.4 is the same as 3.40 (or 3 and forty hundredths). Now ask yourself how you would express "three and forty hundredths" as a fraction. It would, of course, be 3 40/100. Reduce this and it becomes 3 2/5.

Most Common Basic Math Mistakes

Careless Mistakes. Misreading a problem, simple arithmetic mistakes, or other careless errors

Not showing all of the steps. This makes it difficult for the teacher to understand how you got the answer.

Not checking your work or not reviewing their test before turning it in. Never leave the test room early!

Not understanding the problem and solving it with the wrong method.

Not understanding of basic concepts and operations, such as fractions, decimals, and basic algebra.

Not paying attention to the units of measure. Not understanding basic terminology, such as "factor," "product," and "quotient."

Not paying attention to the sign of the answer or confusing the sign.

Not using the correct formula or equation for the problem.

Scientific Notation

Science notation is a very simple and effective way of representing very large numbers in simpler forms. For example, instead of writing out 149,600,000,000 meters, which is the estimated distance from the sun, astronomers could easily write it out as 1.496 x 1011 meters.

Scientific notation expresses numbers in their powers of ten. It can also be used to express simple numbers. For example, using scientific notation, $10 = 10^1$ The exponent "1" tell the number of times to multiply by 10 to get the original number.

$100 = 10^2$
$1000 = 10^3$
$10^0 = 1$

When the exponent is negative, it tells us how many times we need to *divide* by ten to get the original number.

For example, 0.025 = 2.5 x 10

The accepted format of scientific notation or writing numbers on their powers of 10 is $a \times 10^n$

Where a must be between 1 and 10, and n must be an integer.

How to Convert a Number To Scientific Notation

To convert a number to scientific notation, place a decimal after the first number that is not a zero, or, after the first number that between 1 and 9.

After placing the decimal, count the number of places the decimal had to move to get the exponent of 10. If the decimal moves to the left, then the exponent to multiply 10 will be in the positive. If the decimal moves from right to left, it will be a negative power of 10.

For example, to convert 29010, we need to place a decimal after 2, since 2 is the first non zero number. We would then have 2.91

If we were to convert 0.0167, we need to place the decimal after 1, since the first two numbers before 1 are zeros, and do not fall between 1 and 9. We would thus have 1.67

To complete the conversion of 29010 to scientific notation, we would get 2.91×10^{4}

The 10 is raised to the power of 4, because there are 4 places counting from right to left. This scientific notation is positive because the decimal moved to the left.

$0.0167 = 1.67 \times 10^{-2}$

In this example, the decimal place moved from left to right by 2 spaces thus the 10 is raised to the power of 2. It is negative, because the decimal moved to the right.
How to convert from scientific notation

You may also need to convert numbers that are already represented in scientific notation or in their power of ten, to regular numbers.

First it is important to remember these two laws.

If the power is positive, shift decimal to the right
If the power is negative, shift decimal point to the left

Examples

Convert 3.201×10^3

This scientific notation is positive so shift the decimal to the right by 2 spaces, which is the power of the 10. We thus have: $3.201 \times 10^3 = 3201$

Another example

Convert
1.03×10^{-4}
The scientific notation here is negative and so we need to shift decimal to the left. Thus $1.03 \times 10^{-4} = 0.000103$ The decimal was shifted 4 spaces to the left.

Exponents and Radicals

Exponents: Tips, Short-cuts & Tricks

Exponents are just shorthand for saying that you're multiplying a number by itself two or more times.

For instance, instead of saying 5 x 5 x 5, you can show that you're multiplying 5 by itself 3 times if you just write 5^3 .

We usually say this as "five to the third power" or "five to the power of three." In this example, the raised 3 is an "exponent," and the 5 is the "base."

You can even use exponents with fractions. For instance, $1/2^{\,3}$ means you're multiplying 1/2 x 1/2 x 1/2. (The answer is 1/8).

Multiplying Exponents

For exponents with the same base, for instance 5^3 X 5^2 , add the exponents and keep the same base. The answer, then, is 5^5.

If the bases are different, for example, in 5^3 X 3^2, you have to do the math the long way to figure it out.

5 x 5 x 5 = 125, and 3 X 3 = 9.

125 X 9 = 1125

Dividing Exponents

For exponents with the same base, subtract the exponents. In the problem above, 5^3 X 5^2 , 3 - 2 = 1. 5 to the power of 1 is 5.

Here are some Quick things to remember

Any number to the power of 1 is that number.

Any number raised to the power of 0 is 1.

Number (x)	X	X
1	1	1
2	4	8
3	9	27
4	16	64
5	25	125
6	36	216
7	49	343
8	64	512
9	81	729
10	100	1000
11	121	1331
12	144	1728
13	169	2197
14	196	2744
15	225	3375
16	256	4096

Simplifying Radicals

To simplify a radical (square root) Find the largest perfect square which will divide evenly into the number under the square root sign.

$\sqrt{60}$ - this can be re-written as $\sqrt{(3 * 16)}$

and then $3\sqrt{16}$
and $\sqrt{128} = \sqrt{2}\sqrt{64} = 8\sqrt{2}$

Multiply and Dividing Exponents – Video Tutorial

HTTPS://YOUTU.BE/NDJCVEPR6ZMCC

Basic Math Answer Sheet

1. (A) (B) (C) (D)
2. (A) (B) (C) (D)
3. (A) (B) (C) (D)
4. (A) (B) (C) (D)
5. (A) (B) (C) (D)
6. (A) (B) (C) (D)
7. (A) (B) (C) (D)
8. (A) (B) (C) (D)
9. (A) (B) (C) (D)
10. (A) (B) (C) (D)
11. (A) (B) (C) (D)
12. (A) (B) (C) (D)
13. (A) (B) (C) (D)
14. (A) (B) (C) (D)
15. (A) (B) (C) (D)
16. (A) (B) (C) (D)
17. (A) (B) (C) (D)
18. (A) (B) (C) (D)
19. (A) (B) (C) (D)
20. (A) (B) (C) (D)
21. (A) (B) (C) (D)
22. (A) (B) (C) (D)
23. (A) (B) (C) (D)
24. (A) (B) (C) (D)
25. (A) (B) (C) (D)
26. (A) (B) (C) (D)
27. (A) (B) (C) (D)
28. (A) (B) (C) (D)
29. (A) (B) (C) (D)
30. (A) (B) (C) (D)
31. (A) (B) (C) (D)
32. (A) (B) (C) (D)
33. (A) (B) (C) (D)
34. (A) (B) (C) (D)
35. (A) (B) (C) (D)
36. (A) (B) (C) (D)
37. (A) (B) (C) (D)
38. (A) (B) (C) (D)
39. (A) (B) (C) (D)
40. (A) (B) (C) (D)
41. (A) (B) (C) (D)
42. (A) (B) (C) (D)
43. (A) (B) (C) (D)
44. (A) (B) (C) (D)
45. (A) (B) (C) (D)
46. (A) (B) (C) (D)
47. (A) (B) (C) (D)
48. (A) (B) (C) (D)
49. (A) (B) (C) (D)
50. (A) (B) (C) (D)
51. (A) (B) (C) (D)
52. (A) (B) (C) (D)
53. (A) (B) (C) (D)
54. (A) (B) (C) (D)
55. (A) (B) (C) (D)
56. (A) (B) (C) (D)
57. (A) (B) (C) (D)
58. (A) (B) (C) (D)
59. (A) (B) (C) (D)
60. (A) (B) (C) (D)
61. (A) (B) (C) (D)
62. (A) (B) (C) (D)
63. (A) (B) (C) (D)
64. (A) (B) (C) (D)

Basic Math Practice Questions

Fractions, Decimals and Percent

1. 2/3 + 5/12 =

a. 9/17`
b. 3/11
c. 7/12
d. 1 1/12

2. 3/5 + 7/10 =

a. 1 1/10
b. 7/10
c. 1 3/10
d. 1 1/12

3. 4/5 – 2/3 =

a. 2/2
b. 2/13
c. 1
d. 2/15

4. 13/16 – 1/4 =

a. 1
b. 12/12
c. 9/16
d. 7/16

5. 15/16 x 8/9 =

a. 5/6
b. 16/37
c. 2/11
d. 5/7

6. 3/4 x 5/11 =

a. 2/15
b. 15/44
c. 3/19
d. 15/44

7. 5/8 ÷ 2/3 =

a. 15/16
b. 10/24
c. 5/12
d. 1 2/5

8. 2/15 ÷ 4/5 =

a. 6/65
b. 6/75
c. 5/12
d. 1/6

9. In a class of 83 students, 72 are present. What percent of the students are absent? Provide answer up to two significant digits.

a. 12%
b. 13%
c. 14%
d. 15%

10. A woman spent 15% of her income on an item and ends with $120. What percentage of her income is left?

a. 12%
b. 85%
c. 75%
d. 95%

11. X% of 120 = 30. Solve for X.

a. 15
b. 12
c. 4
d. 25

12. Simplify 6 3/5 – 4 4/5

a. 1 4/5
b. 2 3/5
c. 2 9/5
d. 1 1/5

13. Express 25% as a fraction.

a. 1/4
b. 7/40
c. 6/25
d. 8/28

14. Express 125% as a decimal.

a. .125
b. 12.5
c. 1.25
d. 125

15. Express 24/56 as a reduced common fraction.

a. 4/9
b. 4/11
c. 3/7
d. 3/8

16. Express 71/1000 as a decimal.

a. .71
b. .0071
c. .071
d. 7.1

17. .4% of 36 is

a. 1.44
b. .144
c. 14.4
d. 144

18. Express 0.27 + 0.33 as a fraction.

a. 3/6
b. 4/7
c. 3/5
d. 2/7

19. 8 is what percent of 40?

a. 10%
b. 15%
c. 20%
d. 25%

20. 3.14 + 2.73 + 23.7 =

a. 28.57
b. 30.57
c. 29.56
d. 29.57

21. What is 1/3 of 3/4?

a. 1/4
b. 1/3
c. 2/3
d. 3/48

22. 15 is what percent of 200?

a. 7.5%
b. 15%
c. 20%
d. 17.50%

23. A boy has 5 red balls, 3 white balls and 2 yellow balls. What percent of the balls are yellow?

a. 2%
b. 8%
c. 20%
d. 12%

24. Add 10% of 300 to 50% of 20

a. 50
b. 40
c. 60
d. 45

25. Convert 75% to a fraction.

a. 2/100
b. 85/100
c. 3/4
d. 4/7

26. Multiply 3 by 25% of 40.

a. 75
b. 30
c. 68
d. 35

27. What is 10% of 30 multiplied by 75% of 200?

a. 450
b. 750
c. 20
d. 45

28. Convert 4/20 to percent.

a. 25%
b. 20%
c. 40%
d. 30%

29. Write 765.3682 to the nearest 1000th.

a. 765.368
b. 765.36
c. 765.3682
d. 765.3

30. What number is in the ten thousandths place in 1.7389?

a. 1
b. 8
c. 9
d. 3

Exponents, Radicals and Square Root

31. Express in 3^4 standard form

a. 81
b. 27
c. 12
d. 9

32. Simplify $4^3 + 2^4$

a. 45
b. 108
c. 80
d. 48

33. If x = 2 and y = 5, solve $xy^3 - x^3$

a. 240
b. 258
c. 248
d. 242

34. X^3 x X^2

a. 5^x
b. x^{-5}
c. x^{-1}
d. X^5

35. Divide 243 by 3^3

a. 243
b. 11
c. 9
d. 27

36. $7^5 - 3^5$ =

a. 15,000
b. 16,564
c. 15,800
d. 15,007

37. Solve for x if, $10^2 \times 100^2 = 1000^x$

a. $x = 2$

b. $x = 3$

c. $x = -2$

d. $x = 0$

38. Express 9 x 9 x 9 in exponential form and standard form.

a. $9^3 = 719$

b. $9^3 = 629$

c. $9^3 = 729$

d. $10^3 = 729$

39. Multiply 0.27 by 9^2

a. 218.7

b. 21.87

c. 21

d. 20.87

40. Solve $3^8/3^5$

a. 3^3

b. 3^5

c. 3^6

d. 3^4

41. Simply $\sqrt{27} * \sqrt{81}$

a. $\sqrt{3^7}$

b. $\sqrt{3^2}$

c. $\sqrt{3^3}$

d. $\sqrt{5^3}$

42. $\sqrt{8} * 3\sqrt{12}$

a. $12\sqrt{3}$
b. $3\sqrt{6}$
c. $12\sqrt{6}$
d. $8\sqrt{6}$

43. What is the result of the multiplication $(3+\sqrt{5})^{120} * \sqrt{(14-\sqrt{180})^{119}}$ **?**

a. $4^{59}(3+\sqrt{5})$
b. $2^{60}(3-\sqrt{5})$
c. $2^{119}(3+2\sqrt{5})$
d. $2^{238}(3+\sqrt{5})$

44. If $2^{x-1} = 3$**, find the value of** 8^x**.**

a. 16
b. 36
c. 186
d. 216

45. What is the square root of $\sqrt{225}$

a. 25
b. 15
c. 5
d. 13

46. Solve √144

a. 14
b. 72
c. 24
d. 12

47. Solve √121

a. 11
b. 12
c. 21
d. None of the above

48. What is the square root of √36

a. 16
b. 18
c. 6
d. 13

49. Solve √100

a. 100
b. 10
c. 110
d. 50

Order Of Operation

50. 7 + 2 x (6 + 3) ÷ 3 - 7 =

a. 6
b. 5
c. 7
d. 4

51.11 + 19 x 2 =

a. 60
b. 50
c. 49
d. 54

52. (14 + 2) x 2 + 3 =

a. 21
b. 35
c. 80
d. 43

53. 120 ÷ (6 + 12 x 2)

a. 150
b. 40
c. 6
d. 4

54. 12 + 2 x 44

a. 100

b. 616

c. 110

d. 600

55. 10 x 2 – (7 + 9)

a. 21

b. 16

c. 4

d. 13

Answer Key

1. D

A common denominator is needed,which both 3 and 12 will divide into. So, 8 + 5/12 = 13/12 = 1 1/12

2. C

A common denominator is needed for 5 and 10.
6 + 7/10 = 13/10 = 1 3/10

3. D

A common denominator is needed for 5 and 3.
12 - 10/15 = 2/15

4. C

A common denominator is needed for 16 and 4.
13 - 4/16 = 9/16

5. A

Since there are common numerators and denominators to cancel out, we cancel out 15/16 x 8/9 to get 5/2 x 1/3, and then multiply numerators and denominators to get 5/6

6. D

Since there are no common numerators and denominators to cancel out, we simply multiply the numerators and then the denominators. So 3 x 5/4 x 11 = 15/44

7. A

To divide fractions, multiply the first fraction with the inverse of the second. 5/8 x 3/2, = 15/16

8. D

Multiply the first fraction with the inverse of the second.
2/15 x 5/4, (cancel out) = 1/3 x 1/2 = 1/6

9. B
Number of absent students = 83 – 72 = 11
Percentage of absent students is found by proportioning the number of absent students to the total number of students in the class = 11 * 100/83 = 13.25

Checking the answer, round 13.25 to the nearest whole number: 13%.

10. B
Spent 15%, so 100% - 15% = 85%

11. D
X% of 120 = 30,
X/100 = 30/120
So X = 30/120 x 100/1
3000/120 = 300/12
X = 25

12. A
(6-4) (3/5 – 4/5) = 2 (3-4/5) = since 3 is less than 4, we would have to subtract 1 from the whole number besides the fraction, therefore 1 4/5

13. A
25% = 25/100 = 1/4

14. C
125/100 = 1.25

15. C
24/56 = 3/7 (divide numerator and denominator by 8)

16. C
Converting a fraction into a decimal – divide the numerator by the denominator – so 71/1000 = .071. Dividing by 1000 moves the decimal point 3 places to the left.

17. B

.4/100 * 36 = .4 * 36/100 = .144

18. C

To convert a decimal to a fraction, take the places of decimal as your denominator, here, 2, so in 0.27, '7' is in the 100th place, so the fraction is 27/100 and 0.33 becomes 33/100.

Next estimate the answer quickly to eliminate obvious wrong choices. 27/100 is about 1/4 and 33/100 is 1/3. 1/3 is slightly larger than 1/4, and 1/4 + 1/4 is 1/2, so the answer will be slightly larger than 1/2.
Looking at the choices, Choice A can be eliminated since 3/6 = 1/2. Choice D, 2/7 is less than 1/2 and be eliminated. The answer is going to be Choice B or Choice C.

Do the calculation, 0.27 + 0.33 = 0.60 and 0.60 = 60/100 = 3/5, Choice C is correct.

19. C

This is an easy question, and shows how you can solve some questions without doing the calculations. The question is, 8 is what percent of 40. Take easy percentages for an approximate answer and see what you get.

10% is easy to calculate because you can drop the zero, or move the decimal point. 10% of 40 = 4, and 8 = 2 X 4, so, 8 must be 2 X 10% = 20%.

Here are the calculations which confirm the quick approximation.
8/40 = X/100 = 8 * 100 / 40X = 800/40 = X = 20

20. D

3.14 + 2.73 = 5.87 and 5.87 + 23.7 = 29.57

21. A
1/3 X 3/4 = 3/12 = 1/4
To multiply fractions, multiply the numerator and denominator.

22. A
15/200 = X/100
200X = (15 * 100)
1500/200 Cancel zeros in the numerator and denominator
15/2 = 7.5%.

Notice that the questions asks, What 15 is what percent of 200? The question does *not* ask, what is 15% of 200! The answers are very different.

23. C
Total no. of balls = 10, no. of yellow balls = 2, answer = 2/10 X 100 = 20%.

24. B
10% of 300 = 30 and 50% of 20 = 10 so 30 + 10 = 40.

25. C
75% = 75/100 = 3/4

26. B
25% of 40 = 10 and 10 x 3 = 30

27. A
10% of 30 = 3 and 75% of 200 = 150, 3 X 150 = 450

28. B
4/20 X 100 = 1/5 X 100 = 20%

29. A
The number is 51.738. The last digit, in the 1,000th place, 2, is less than 5, so it is discarded. Answer = 765.368.

30. C
9 is in the ten thousandths place in 1.7389, which is 4 places to the right of the decimal point.

Exponents, Radicals and Square Root

31. A
3 x 3 x 3 x 3 = 81

32. C
(4 x 4 x 4) + (2 x 2 x 2 x 2) = 64 + 16 = 80

33. D
$2(5)^3 - (2)^3$ = 2(125) – 8 = 250 – 8 = 242

34. D
$X^3 \times X^2 = X^{3+2} = X^5$

35. C
$243/3^3$ 3 x 3 x 3 = 27
243/27 = 9

36. B
(7 x 7 x 7 x 7 x 7) - (3 x 3 x 3 x 3 x 3) = 16,807 – 243 = 16,564

37. A
10 x 10 x 100 x 100 = 1000^x, =100 x 10,000 = 1000^x, = 1,000,000 = 1000x = x = 2

38. C
Exponential form is 9^3 and standard from is 729

39. B

0.27 (9 x 9) = 0.27 x 81 = 21.87

40. A

$3^{8-5} = 3^3$

To divide exponents with the same base, subtract the exponents.

41. A

$\sqrt{27} * \sqrt{81} = \sqrt{3^3} * \sqrt{3^4}$

$\sqrt{3^{3+4}} = \sqrt{3^7}$

42. C

$\sqrt{8} * 3\sqrt{12} = (\sqrt{2} * \sqrt{4}) * 3(\sqrt{4} * \sqrt{3})$

$= 2\sqrt{2} * ((3 * 2) * \sqrt{3})$

$= 2\sqrt{2} * 6 * \sqrt{3}$

$= 12 \sqrt{2} * \sqrt{3}$

$= 12 \sqrt{6}$

43. D

First, we need to simplify the term inside the root:

$\sqrt{180} = \sqrt{(62.5)} = 6\sqrt{5}$

$(3 + \sqrt{5})^{120} . \sqrt{(14 - \sqrt{180})^{119}} = (3 + \sqrt{5})^{120} * \sqrt{(14 - 6\sqrt{5})119}$

Notice that we need to find the way to simplify the second term of the multiplication. $14 - 6\sqrt{5}$ is a perfect square:

$- 6\sqrt{5} = - 2ab$

$a = \sqrt{5}$ and $b = 3$

$14 = a^2 + b^2 = (\sqrt{5})^2 + 3^2$. So;

$(3 + \sqrt{5})120 * \sqrt{(14 - 6\sqrt{5})^{119}} = (3 + \sqrt{5})120 * \sqrt{((3 - \sqrt{5})2)^{119}}$

$= (3 + \sqrt{5})^{120} * (3 - \sqrt{5})\ 2^{119} / 2$

$= (3 + \sqrt{5})^{120} * (3 - \sqrt{5})^{119}$
Remember that $(a - b)(a + b) = a^2 - b^2$:

$= (3 + \sqrt{5}) * (3 + \sqrt{5})^{119} * (3 - \sqrt{5})^{119}$
$= (3 + \sqrt{5}) * (32 - (\sqrt{5})2)^{119}$
$= (3 + \sqrt{5}) * 4^{119}$
$= 2^{238}(3 + \sqrt{5})$

44. D
In this question, we do not need to try to find the value of x. Notice that the numbers containing x as power are of base 2 both in the given and asked expressions. So, let us find the value of 2^x first:

$2^{x-1} = 3$

$2^{-1} * 2^x = 3$

$2^x = 3 * 2$

$2^x = 6$

The value of 8x is asked:

$8x = (2^3)^x = 2^{3x} = 6^3 = 216$

45. B
$\sqrt{225} = 15$

46. D
$\sqrt{144} = 12$

47. A
$\sqrt{121} = 11$

48. C
$\sqrt{36} = 6$

49. B
$\sqrt{100} = 10$

Order of Operation

50. A

7 + 2 x (6 + 3) ÷ 3 - 7 =

7 + 2 x (9) ÷ 3 - 7 Brackets first

7 + 2 x (9) ÷ 3 - 7 left to right multiplication and division

7 + ((2 x 9) ÷ 3) - 7

= 7 + (18 ÷ 3) - 7

= 7 + 6 - 7 Now left to right addition and subtraction

= 13 - 7 = 6

51. C
11 + 19 x 2 first, left to right multiplication

11 + 38

= 49

52. B
(14 + 2) x 2 + 3 = Operations inside parenthesis first

16 X 2 + 3 Next left to right multiplication

32 + 3 = 35

53. D

120 ÷ (6 + 12 x 2) Operations inside parenthesis first - multiplication before addition

120 ÷ (6 + 24) Now addition in parenthesis

120 ÷ 30 = 4

54. A

12 + 2 x 44 Multiplication first

12 + 88 = 100

55. C

10 x 2 – (7 + 9) Operations inside parenthesis first

10 x 2 – 16 Next, left to right, multiplication

20 - 16 = 4

Data, Statistics and Probability

The data interpretation portion of the HiSET© covers:

- Interpret data in a graph or chart
- Graph data
- Calculate mean, median and mode
- Identify dependent and independent variables

Mode, Mean and Median

The Mode, Mean and Median, are types of averages.

The mean is the average calculated by adding the numbers and dividing by the number of items in the data set. The median is the middle value in a data set. To calculate the median, put the numbers in order, and the median will be the middle number. If there is an even number of items in the data set, then the median is found by taking the mean (average) of the two middlemost numbers. See our example below. The mode is the most frequently occurring number. If no number is repeated, then there is no mode.

Examples

Find the median, mode and mean of the following list:
6, 7, 8, 12, 14, 6, 7, 10

Find the mean

First add the numbers
6 + 7 + 8 + 12 + 14 + 6 + 7 + 10 = 70
There are 8 numbers in the list, so divide by 9
70/8 = 8.75 = mean

Find the median

First put the numbers in order
6, 6, 7, 7, 8, 10, 12, 14,
The data set has an even number of numbers, so the median is the average of 7 and 8. (7 + 8)/2 = 7.5

Find the mode

The mode is the most frequently occurring number. Here 6 and 7 both occur twice, so they are both considered the mode.

Dependent and Independent Variables

Independent variables are exactly just what you would think - a variable that is "independent" and therefore isn't changed by the other variables. For example, someone's age might be an independent variable. Other factors (such as what they eat or what school they go to) has no effect on their age.

When setting up a test, you are trying to see of the independent variable has an effect on the other (dependent) variable.

A *dependent* variable is just what you would think. It is

something that depends on other factors. For example, a test score could be a dependent variable because it could change depending on several factors such as how much you studied, how much sleep you got the night before you took the test, or even how hungry you were when you took it. Usually when you are looking for a relationship between two things you are trying to find out what makes the dependent variable change the way it does.

Concepts in Statistics and Probability

Probability Concepts
Probability is the study of chance and randomness.

Experiment, Sample Space, and Event

Experiment: Any process that yields a result (e.g., tossing a coin).
Sample Space (S): The set of all possible outcomes.
Event: A specific set of outcomes of interest.

Probability Rules
Probability of an event is between 0 and 1.
The sum of probabilities for all possible outcomes is 1.

Addition Rule: Probability of A or B = P(A) + P(B) - P(A and B).
Multiplication Rule: Probability of A and B (if independent) = P(A) × P(B).

Conditional Probability and Independence

Conditional Probability: The probability of event A given event B has occurred, denoted P(A|B).

Independence: Two events are independent if the occurrence of one does not affect the probability of the other.

Random Variables
Discrete Random Variable: Takes countable values (e.g., number of heads in coin tosses).
Continuous Random Variable: Takes any value in a continuum (e.g., height, weight).
Probability Distribution: Lists possible values and their probabilities (for discrete) or the probability density function (for continuous).

Expected Value and Variance
Expected Value (Mean): The long-run average outcome weighted by probabilities.
Variance: The expected squared deviation from the mean.

Answer Sheet

	A	B	C	D
1	○	○	○	○
2	○	○	○	○
3	○	○	○	○
4	○	○	○	○
5	○	○	○	○
6	○	○	○	○
7	○	○	○	○
8	○	○	○	○
9	○	○	○	○
10	○	○	○	○
11	○	○	○	○
12	○	○	○	○
13	○	○	○	○
14	○	○	○	○
15	○	○	○	○
16	○	○	○	○
17	○	○	○	○
18	○	○	○	○
19	○	○	○	○
20	○	○	○	○

Data Analysis and Statistics Practice Questions

1. Find the median of the set of numbers: 1,2,3,4,5,6,7,8,9 and 10.

a. 55
b. 10
c. 1
d. 5.5

2. Find the median of the set of numbers: 21, 3, 7, 17, 19, 31, 46, 20 and 43.

a. 19
b. 20
c. 3
d. 167

3. Find the median of the set of numbers: 100, 200, 450, 29, 1029, 300 and 2001.

a. 300
b. 29
c. 7
d. 4,080

4. The following represents age distribution of students in an elementary class. Find the mode of the values: 7, 9, 10, 13, 11, 7, 9, 19, 12, 11, 9, 7, 9, 10, 11.

a. 7
b. 9
c. 10
d. 11

5. Find the mode from these test results: 90, 80, 77, 86, 90, 91, 77, 66, 69, 65, 43, 65, 75, 43, 90.

a. 43
b. 77
c. 65
d. 90

6. Find the mode from these test results: 17, 19, 18, 17, 18, 19, 11, 17, 16, 19, 15, 15, 15, 17, 13, 11.

a. 15
b. 11
c. 17
d. 19

7. Find the mean of these set of numbers: 100, 1050, 320, 600 and 150.

a. 333
b. 444
c. 440
d. 320

8. The following numbers represent the ages of people on a bus: 3, 6, 27, 13, 6, 8, 12, 20, 5, 10. Calculate their mean of their ages.

a. 11

b. 6

c. 9

d. 110

9. The following are the number of people who attend church every Friday for 7 weeks: 62, 18, 39, 13, 16, 37, 25. Find the mean.

a. 25

b. 210

c. 62

d. 30

10. What number would you divide by to calculate the mean of 6, 8, 9, and 11?

a. 2

b. 3

c. 4

d. 5

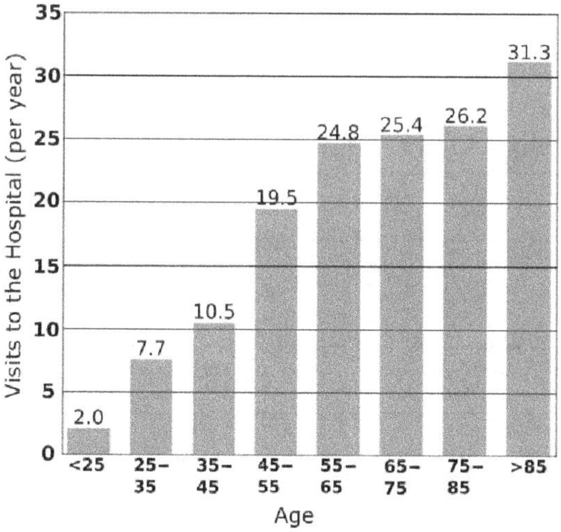

11. Consider the graph above.

How many hospital visits per year does a person aged 85 or more make?

a. 26.2

b. 31.3

c. More than 31.3

d. A decision cannot be made from this graph.

12. Based on this graph, how many visits per year do you expect a person that is 95 or older to make?

a. 31.3 or more

b. Less than 31.3

c. 31.3

d. A decision cannot be made from this graph.

13. Consider the following population growth chart.

Country	Population 2000	Population 2005
Japan	122,251,000	128,057,000
China	1,145,195,000	1,341,335,000
United States	253,339,000	310,384,000
Indonesia	184,346,000	239,871,000

What country is growing the fastest?

a. Japan

b. China

c. United States

d. Indonesia

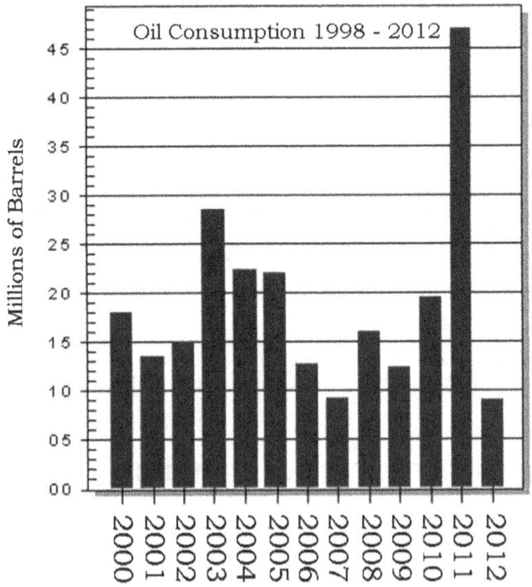

14. The graph above shows oil consumption in millions of barrels for the period, 1998 - 2012. What year did oil consumption peak?

a. 2011
b. 2010
c. 2008
d. 2009

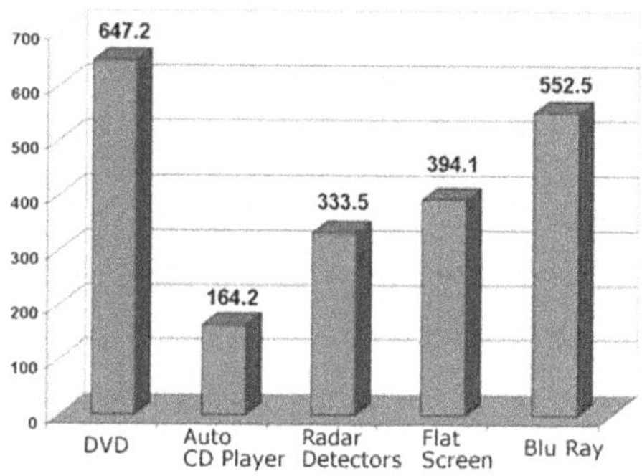

15. Consider the graph above. What is the third best-selling product?

a. Radar Detectors
b. Flat Screen
c. Blu Ray
d. Auto CD Players

16. Which two products are the closest in the number of sales?

a. Blu Ray and Flat Screen TV
b. Flat Screen TV and Radar Detectors
c. Radar Detectors and Auto CD Players
d. DVD players and Blu Ray

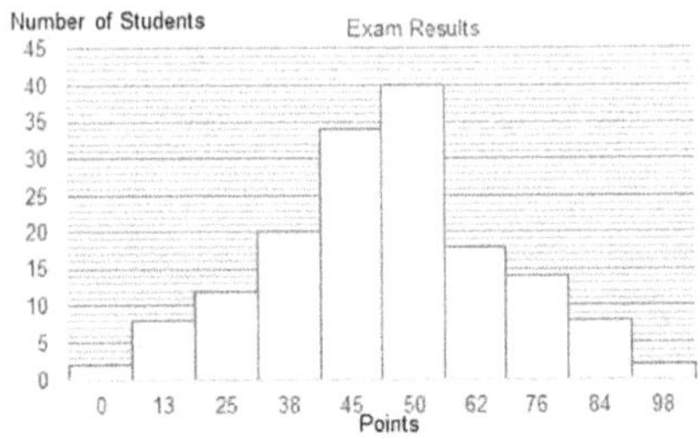

17. According to the graph above, how many students received 62 or higher on the exam?

a. 40

b. 42

c. 48

d. 52

18. Mr. Jones needs to order enough textbooks (T) for all the students (S) that register in his course.

Which is the dependent variable?

a. T is the dependent variable

b. S is the dependent variable

c. We cannot tell from this description

19. A waitress needs bring menus (M) for each person (P) at a table. Which variable is the independent variable?

a. M is the independent variable

b. P is the independent variable

c. We cannot tell from this description

20. Graph the following data

Salt (grams)	Water (grams)
30	90
45	135

Answer Key

1. D

First arrange the numbers in a numerical sequence – 1,2,3,4,5,6,7,8,9, 10. Then find the middle number or numbers. The middle numbers are 5 and 6. The median = 5 + 6/2 = 11/2 = 5.5

2. B

First arrange the numbers in a numerical sequence – 3,7, 17, 19, 20, 21, 31, 43, 46. Next find the middle number. The median = 20

3. A

First arrange the numbers in a numerical sequence – 29,100, 200, 300, 450, 1029, 2001. Next find the middle number. The median = 300

4. B

Simply find the most recurring number. The most occurring number in the series is 9

5. D

Simply find the most recurring number. The most occurring number in the series is 90.

6. C

Simply find the most recurring number. The most occurring number in the series is 15.

7. B

First add all the numbers 100 + 1050 + 320 + 600 + 150 = 2220. Then divide by 5 (the number of data provided) = 2220/5 = 444

8. A

First add all the numbers 3 + 6 + 27 + 13 + 6 + 8 + 12 + 20 + 5 + 10 = 110. Then divide by 10 (the number of data provided) = 110/10 = 11

9. D

First add all the numbers 62 + 18 + 39 + 13 + 16 + 37 + 25 = 210. Then divide by 7 (the number of data provided) = 210/7 = 30

10. C

To calculate the mean of 6, 8, 9, and 11, divide by 4.

11. A

Based on this graph, a person that is 85 or older will make 26.2 visits to the hospital every year.

12. A

A person aged 95 or older would make 31.3 or more visits.

13. D

Indonesia is growing the fastest, at about 30%.

14. A

According to the graph, oil consumption peaked in 2011.

15. B

Flat Screen TV are the third best-selling product.

16. B

The two products that are closest in the number of sales, are Flat Screen TVs and Radar Detectors.

17. B

42 students received a score of 62 or higher.

18. A

The number of textbooks required will depend on the number of students, so T is the dependent variable.

19. B

The number of people at the table (P) is the independent variable.

20. A graph of the data provided will look like,

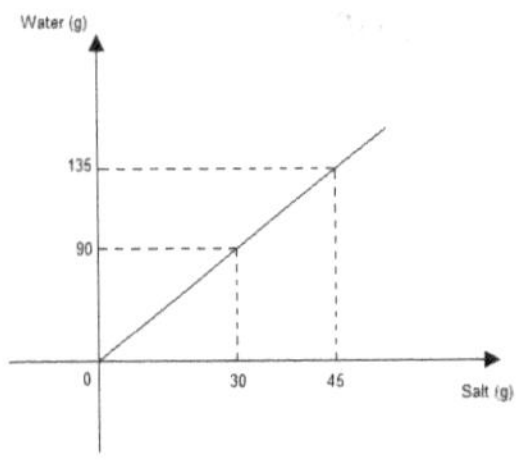

Word Problems

MOST STUDENTS FIND WORD PROBLEMS DIFFICULTY SO WE HAVE INCLUDED AN EXTRA CHAPTER WITH A TUTORIAL AND EXTRA PRACTICE.

How to Solve Word Problems

Do you know what the biggest tip for solving word problems is?

Practice regularly and systematically.

Sounds simple and easy right? Yes it is, and yes it really does work.

Word problems are a way of thinking and require you to translate a real-world problem into mathematical terms.

Some math teachers say that learning how to think mathematically is the main reason for teaching word problems.

So what does that mean?

Studying word problems and math in general requires a logical and mathematical frame of mind. The only way you can get this is by practicing regularly, which means every day.

It is critical that you practice word problems every day for the 5 days before the exam as the absolute minimum.

If you practice and miss a day, you have lost the mathematical frame of mind and the benefit of your previous practice is gone. You must start all over again.

Everything is important.

All the information given in the problem has some purpose. There is no unnecessary information! Word problems are typically around 50 words in 2 or 3 sentences.

Often, the relationships are complicated. To explain everything, every word counts.

Make sure that you use every piece of information.

9 steps to solving word problems

Step 1 – Read through the problem at least three times. The first reading should be a quick scan, and the next two readings should be done slowly to find answers to these important questions:

What does the problem ask? (Usually located towards the end of the problem)

What does the problem imply? (This is usually a point you were asked to remember).

Mark all information, and underline all important words or phrases.

Step 2 – Try to make a pictorial representation of the problem such as a circle and an arrow to show travel. This makes the problem a bit more real and sensible to you.

A favorite word problem is something like, 1 train leaves Station A traveling at 100 km/hr and another train leaves Station B traveling at 60 km/hr. ...

Draw a line, the two stations, and the two trains at either end. This will help clarify the situation in your mind.

Step 3 – Use the information you have to make a table with a blank portion to show information you do not know.

Step 4 – Assign a single letter to represent each unknown data in your table. You can write down the unknown that each letter represents so that you do not make the error of assigning answers to the wrong unknown, because a word problem may have multiple unknowns and you will need to create equations for each unknown.

Step 5 – Translate the English terms in the word problem into a mathematical algebraic equation. Remember that the main problem with word problems is that they are not expressed in regular math equations. You ability to identify correctly the variables and translate the word problem into an equation determines your ability to solve the problem.

Step 6 – Check the equation to see if it looks like regular equations that you have seen before, and whether it looks sensible. Does the equation appear to represent the information in the question? Take note that you may need to rewrite some formulas needed to solve the word problem equation. For example, word distance problems may need rewriting the distance formula, which is Distance = Time x Rate. If the word problem requires that you solve for time you will need to use Distance/Rate and Distance/Time to solve for Rate. If you understand the distance word problem you should be able to identify the variable you need to solve for.

Step 7 – Use algebra rules to solve the derived equation. Take note that the laws of equation demands that what is done on this side of the equation has to also be done on the other side. You have to solve the equation so that the unknown ends up alone on one side. Where there are multiple unknowns you will need to use elimination or substitution methods to resolve all the equations.

Step 8 – Check your final answers to see if they make sense with the information given in the problem. For example if the word problem involves a discount, the final price should be less or if a product was taxed then the final answer has to cost more.

Step 9 – Cross check your answers by placing the answer or answers in the first equation to replace the unknown or unknowns. If your answer is correct then both side of the equation must equate or equal. If your answer is not correct then you may have derived a wrong equation or solved the equation wrongly. Repeat the necessary steps to correct.

Types of Word Problems

Word problems can be classified into 12 types. Below are examples of each type with a complete solution. Some types of word problems can be solved quickly using multiple choice strategies and some cannot. Always look for ways to estimate the answer and then eliminate choices.

.1. Age

A girl is 10 years older than her brother. By next year, she will be twice the age of her brother. What are their ages now?

a. 25, 15
b. 19, 9
c. 21, 11
d. 29, 19

Solution: B

We will assume that the girl's age is "a" and her brother's age is "b." This means that based on the information in the first sentence,
$a = 10 + b$

Next year, she will be twice her brother's age, which gives, $a + 1 = 2(b + 1)$

We need to solve for one unknown factor and then use the answer to solve for the other. To do this we substitute the value of "a" from the first equation into the second equation. This gives

$10+b + 1 = 2b + 2$
$11 + b = 2b + 2$
$11 – 2 = 2b – b$
$b= 9$

$9 = b$ this means that her brother is 9 years old. Solving for the girl's age in the first equation gives $a = 10 + 9$. $a = 19$ the girl is aged 19. So, the girl is aged 19 and the boy is 9

2. Distance or Speed

Two boats travel down a river towards the same destination, starting at the same time. One is traveling at 52 km/hr, and the other boat at 43 km/hr. How far apart will they be after 40 minutes?

a. 46.67 km
b. 19.23 km
c. 6.4 km
d. 14.39 km

Solution: C

After 40 minutes, the first boat will have traveled = 52 km/hr x 40 minutes/60 minutes = 34.7 km
After 40 minutes, the second boat will have traveled = 43 km/hr x 40/60 minutes = 28.66 km
Difference between the two boats will be 34.7 km – 28.66 km = 6.04 km.

Multiple Choice Strategy

First estimate the answer. The first boat is traveling 9 km. faster than the second, for 40 minutes, which is 2/3 of an hour. 2/3 of 9 = 6, as a rough guess of the distance apart.

Choices A, B and D can be eliminated right away.

3. Ratio

The instructions in a cookbook state that 700 grams of flour must be mixed in 100 ml of water, and 0.90 grams of salt added. A cook however has just 325 grams of flour. What is the quantity of water and salt that he should use?

a. 0.41 grams and 46.4 ml
b. 0.45 grams and 49.3 ml
c. 0.39 grams and 39.8 ml
d. 0.25 grams and 40.1 ml

Solution: A

The Cookbook states 700 grams of flour, but the cook only has 325. The first step is to determine the percentage of flour he has 325/700 x 100 = 46.4%
That means that 46.4% of all other items must also be used.
46.4% of 100 = 46.4 ml of water
46.4% of 0.90 = 0.41 grams of salt.

Multiple Choice Strategy

The recipe calls for 700 grams of flour but the cook only has 325, which is just less than half, the quantity of water and salt are going to be about half.
Choices C and D can be eliminated right away. Choice B is very close so be careful. Looking closely at choice B, it is exactly half, and since 325 is slightly less than half of 700, it can't be correct.

Choice A is correct.

4. Percent

An agent received $6,685 as his commission for selling a property. If his commission was 13% of the selling price, how much was the property?

a. $68,825
b. $121,850
c. $49,025
d. $51,423

Solution: D

Let's assume that the property price is x. That means from the information given, 13% of x = 6,685
Solve for x,

x = 6685 x 100/13 = $51,423

Multiple Choice Strategy

The commission,13%, is just over 10%, which is easier to work with. Round up $6685 to $6700, and multiple by 10 for an approximate answer. 10 X 6700 = $67,000. You can do this in your head. Choice B is much too big and can be eliminated. Choice C is too small and can be eliminated. Choices A and D are left and good possibilities.

Do the calculations to make the final choice.

5. Sales & Profit

A store owner buys merchandise for $21,045. He transports them for $3,905 and pays his staff $1,450 to stock the merchandise on his shelves. If he does not incur further costs, how much does he need to sell the items to make $5,000 profit?

a. $32,500
b. $29,350
c. $32,400
d. $31,400

Solution: D

Total cost of the items is $21,045 + $3,905 + $1,450 = $26,400

Total cost is now $26,400 + $5000 profit = $31,400

Multiple Choice Strategy

Round off and add the numbers up in your head quickly. 21,000 + 4,000 + 1500 = 26500. Add in 5000 profit for a total of 31500.

Choice B is too small and can be eliminated. Choice C and Choice A are too large and can be eliminated.

6. Tax/Income

A woman earns $42,000 per month and pays 5% tax on her monthly income. If the Government increases her monthly taxes by $1,500, what is her income after tax?

a. $38,400
b. $36,050
c. $40,500
d. $39, 500

Solution: A

Initial tax on income was 5/100 x 42,000 = $2,100
$1,500 was added to the tax to give $2,100 + 1,500 = $3,600
Income after tax is $42,000 - $3,600 = $38,400

7. Simple Interest Word Problems

Simple interest is one type of interest problems. There are always four variables of any simple interest equation. With simple interest, you would be given three of these variables and be asked to solve for one unknown variable. With more complex interest problems, you would have to solve for multiple variables.

The four variables of simple interest are:

P – Principal which refers to the original amount of money put in the account
I – Interest or the amount of money earned as interest
r – Rate or interest rate. This MUST ALWAYS be in decimal format and not in percentage
t – Time or the amount of time the money is kept in the account to earn interest

The formula for simple interest is I = P x r x t

Example 1

A customer deposits $1,000 in a savings account with a bank that offers 2% interest. How much interest will be earned after 4 years?

For this problem, there are 3 variables as expected.

P = $1,000
t = 4 years
r = 2%
I = ?
Before we can begin solving for I using the simple interest formula, we need to first convert the rate from percentage to decimal.

2% = 2/100 = 0.02
Now we can use the formula: I = P x r x t

I = 1,000 x 0.02 x 4 = 80
This means that the $1,000 would have earned an interest of $80 after 4 years. The total in the account after 4 years will thus be principal + interest earned, or 1,000 + 80 = $1,080

Example 2

Sandra deposits $1400 in a savings account with a bank at 5% interest. How long will she have to leave the money in the bank to earn $420 as interest to buy a second-hand car?

In this example, the given information is:
I = $420
P = $1,400
r - 5%
t - ?
As usual, first we convert the rate from percentage to decimal

5% = 5/100 = 0.05

Next, we plug in the variables we know into the simple interest formula - I = P x r x t

420 = 1,400 x 0.05 x t
420 = 70 x t
420 = 70t
t = 420/70
t = 6
Sandra will have to leave her $1,400 in the bank for 6 years to earn her an interest of $420 at a rate of 5%.

Other important simple interest formula to remember are below. To use these formula, do not convert r (rate) to decimal.

P = 100 x interest/ r x t
r = 100 x interest/p x t
t = 100 x interest/ p x r

8. Averaging

The average weight of 10 books is 54 grams. 2 more books were added and the average weight became 55.4. If one of the 2 new books added weighed 62.8 g, what is the weight of the other?

a. 44.7 g
b. 67.4 g
c. 62 g
d. 52 g

Solution: C

Total weight of 10 books with average 54 grams will be = 1 0 × 54 = 540 g

Total weight of 12 books with average 55.4 will be
= 55.4 × 12 = 664.8 g

Total weight of the remaining 2 will be
= 664.8 – 540 = 124.8 g

If one weighs 62.8, the weight of the other will be
= 124.8 g – 62.8 g = 62 g

Multiple Choice Strategy

Averaging problems can be estimated by looking at which direction the average goes. If additional items are added and the average goes up, the new items much be greater than the average. If the average goes down after new items are added, the new items must be less than the average.

Here, the average is 54 grams and 2 books are added which increases the average to 55.4, so the new books must weight more than 54 grams.
Choices A and D can be eliminated right away.

9. Probability

A bag contains 15 marbles of various colors. If 3 marbles are white, 5 are red and the rest are black, what is the probability of randomly picking out a black marble from the bag?

a. 7/15
b. 3/15
c. 1/5
d. 4/15

Solution: A

Total marbles = 15

Number of black marbles = 15 – (3 + 5) = 7
Probability of picking out a black marble = 7/15

10. Two Variables

A company paid a total of $2850 to book for 6 single rooms and 4 double rooms in an hotel for one night. Another company paid $3185 to book for 13 single rooms for one night in the same hotel. What is the cost for single and double rooms in that hotel?

a. single= $250 and double = $345
b. single= $254 and double = $350
c. single = $245 and double = $305
d. single = $245 and double = $345

Solution: D

We can determine the price of single rooms from the information given of the second company. 13 single rooms = 3185.

One single room = 3185 / 13 = 245

The first company paid for 6 single rooms at $245. 245 x 6 = $1470

Total amount paid for 4 double rooms by first company = $2850 - $1470 = $1380

Cost per double room = 1380 / 4 = $345

11. Geometry

The length of a rectangle is 5 in. more than its width. The perimeter of the rectangle is 26 in. What is the width and length of the rectangle?

a. width = 6 inches, Length = 9 inches
b. width = 4 inches, Length = 9 inches
c. width =4 inches, Length = 5 inches
d. width = 6 inches, Length = 11 inches

Solution: B

Formula for perimeter of a rectangle is 2(L + W)
p = 26, so 2(L + W) = p

The length is 5 inches more than the width, so
2(w + 5) + 2w = 26
2w + 10 + 2w = 26
2w + 2w = 26 - 10
4w = 16

W = 16/4 = 4 inches

L is 5 inches more than w, so L = 5 + 4 = 9 inches.

12. Totals and fractions

A basket contains 125 oranges, mangoes and apples. If 3/5 of the fruits in the basket are mangoes and only 2/5 of the mangoes are ripe, how many ripe mangoes are there in the basket?

a. 30
b. 68
c. 55
d. 47

Solution: A
Number of mangoes in the basket is 3/5 x 125 = 75
Number of ripe mangoes = 2/5 x 75 = 30

Word Problem Practice with Video Solutions

HTTPS://YOUTU.BE/6XWA6FO6YCE

Most Common Word Problem Mistakes

Not reading the problem carefully and thoroughly, so that you either misunderstand or solve the problem incorrectly.

Not identifying the important information in the problem, such as the quantities, units, and the operation to be performed.

Not translating the information in the problem into mathematical language and equations.

Not checking the units of measure and making sure they match your final answer.

Not double-checking the answer to ensure it makes sense.

Not understanding the underlying mathematical concept or operation the problem is asking for.

Not using estimation or approximations as a tool to check the reasonableness of your answer.

Proven Strategies for Solving Word Problems

Read Carefully & Read Again
Slow down, re-read, and pinpoint exactly what the problem is asking.

C.U.B.E.S Method

Circle numbers

Underline the question

Circle important words (total, difference, per)

Eliminate extra info

Solve the problem

Translate into Math
Translate words to operations (tota" = +, difference = −, product = ×).

Define your Variables
Label unknowns with meaningful variable names to keep everything straight, "t = time in hours" or "h = height in cm."

Translate Step-by-Step

Break the problem into single statements and convert to equations—then combine. Watch details!

Draw a Diagram or Visualize

Draw the problem (for example rectangles for area problems) or create charts/tables to visualize data.

Plug Numbers in

Plug in a answer choices and see if the result makes sense.

Check Your Work

Good advice for any and all questions! Does the answer make sense? Confirm units and realistic results, and make sure your answer directly answers the question.

Answer Sheet

	A	B	C	D	E		A	B	C	D	E
1	○	○	○	○	○	21	○	○	○	○	○
2	○	○	○	○	○	22	○	○	○	○	○
3	○	○	○	○	○	23	○	○	○	○	○
4	○	○	○	○	○	24	○	○	○	○	○
5	○	○	○	○	○	25	○	○	○	○	○
6	○	○	○	○	○						
7	○	○	○	○	○						
8	○	○	○	○	○						
9	○	○	○	○	○						
10	○	○	○	○	○						
11	○	○	○	○	○						
12	○	○	○	○	○						
13	○	○	○	○	○						
14	○	○	○	○	○						
15	○	○	○	○	○						
16	○	○	○	○	○						
17	○	○	○	○	○						
18	○	○	○	○	○						
19	○	○	○	○	○						
20	○	○	○	○	○						

Word Problem Practice

1. Translate the following into an equation: Five greater than 3 times a number.

a. 3X + 5
b. 5X + 3
c. (5 + 3)X
d. 5(3 + X)

2. Translate the following into an equation: three plus a number times 7 equals 42.

a. 7(3 + X) = 42
b. 3(X + 7) = 42
c. 3X + 7 = 42
d. (3 + 7)X = 42

3. Translate the following into an equation: 2 + a number divided by 7.

a. (2 + X)/7
b. (7 + X)/2
c. (2 + 7)/X
d. 2/(7 + X)

4. Translate the following into an equation: six times a number plus five.

a. 6X + 5
b. 6(X+5)
c. 5X + 6
d. (6 * 5) + 5

5. A box contains 7 black pencils and 28 blue ones. What is the ratio between the black and blue pens?

a. 1:4

b. 2:7

c. 1:8

d. 1:9

6. The manager of a weaving factory estimates that if 10 machines run at 100% efficiency for 8 hours, they will produce 1450 meters of cloth. Due to some technical problems, 4 machines run of 95% efficiency and the remaining 6 at 90% efficiency. How many meters of cloth can these machines will produce in 8 hours?

a. 1334 meters

b. 1310 meters

c. 1300 meters

d. 1285 meters

7. In a local election at polling station A, 945 voters cast their vote out of 1270 registered voters. At polling station B, 860 cast their vote out of 1050 registered voters and at station C, 1210 cast their vote out of 1440 registered voters. What is the total turnout from all three polling stations?

a. 70%

b. 74%

c. 76%

d. 80%

8. If Lynn can type a page in p minutes, what portion of the page can she do in 5 minutes?

a. p/5

b. p - 5

c. p + 5

d. 5/p

9. If Sally can paint a house in 4 hours, and John can paint the same house in 6 hours, how long will it take for both to paint a house?

a. 2 hours and 24 minutes

b. 3 hours and 12 minutes

c. 3 hours and 44 minutes

d. 4 hours and 10 minutes

10. Employees of a discount appliance store receive an additional 20% off the lowest price on any item. If an employee purchases a dishwasher during a 15% off sale, how much will he pay if the dishwasher originally cost $450?

a. $280.90

b. $287.00

c. $292.50

d. $306.00

11. The sale price of a car is $12,590, which is 20% off the original price. What is the original price?

a. $14,310.40

b. $14,990.90

c. $15,108.00

d. $15,737.50

12. Richard gives 's' amount of salary to each of his 'n' employees weekly. If he has 'x' amount of money, how many days he can employ these 'n' employees.

a. sx/7n
b. 7x/nx
c. nx/7s
d. 7x/ns

13. A distributor purchased 550 kilograms of potatoes for $165. He distributed these at a rate of $6.4 per 20 kilograms to 15 shops, $3.4 per 10 kilograms to 12 shops and the remainder at $1.8 per 5 kilograms. If his total distribution cost is $10, what will his profit be?

a. $10.40
b. $8.60
c. $14.90
d. $23.40

14. How much pay does Mr. Johnson receive if he gives half of his pay to his family, $250 to his landlord, and has exactly 3/7 of his pay left over?

a. $3600
b. $3500
c. $2800
d. $1750

15. The cost of waterproofing canvas is .50 a square yard. What's the total cost for waterproofing a canvas truck cover that is 15' x 24'?

a. $18.00

b. $6.67

c. $180.00

d. $20.00

16. The price of a book went from $20 to $25. What percent did the price increase?

a. 5%

b. 10%

c. 20%

d. 25%

17. In the time required to serve 43 customers, a server breaks 2 glasses and slips 5 times. The next day, the same server breaks 10 glasses. Assuming that glasses broken is proportional to customers served, how many customers did she serve?

a. 25

b. 43

c. 86

d. 215

18. A square lawn has an area of 62,500 square meters. What is the cost of building fence around it at a rate of $5.5 per meter?

a. $4000

b. $4500

c. $5000

d. $5500

19. Susan wants to buy a leather jacket that costs $545.00 and is on sale for 10% off. What is the approximate cost?

a. $525

b. $450

c. $475

d. $500

20. Sarah weighs 25 pounds more than Tony. If together they weigh 205 pounds, how much does Sarah weigh in kilograms? Assume 1 pound = 0.4535 kilograms.

a. 41

b. 48

c. 50

d. 52

21. A man buys an item for $420 and has a balance of $3000.00. How much did he have before his purchase?

a. $2,580

b. $3,420

c. $2,420

d. $342

22. The average weight of 13 students in a class of 15 (two were absent that day) is 42 kg. When the remaining two are weighed, the average became 42.7 kg. If one of the remaining students weighs 48 kg., how much does the other weigh?

a. 44.7 kg.
b. 45.6 kg.
c. 46.5 kg.
d. 47.4 kg.

23. The total expense of building a fence around a square-shaped field is $2000 at a rate of $5 per meter. What is the length of one side?

a. 40 meters
b. 80 meters
c. 100 meters
d. 320 meters

24. There were some oranges in a basket. By adding 8/5 of the total to the basket, the new total is 130. How many oranges were in the basket?

a. 60
b. 50
c. 40
d. 35

25. A person earns $25,000 per month and pays $9,000 income tax per year. The Government increased income tax by 0.5% per month and his monthly earning was increased $11,000. How much more income tax will he pay per month?

a. $1260

b. $1050

c. $750

d. $510

Answer Key

Part 1 - Equation Translation

1. A
Five greater than 3 times a number.
5 + 3 times a number.
3X + 5

2. A
Three plus a number times 7 equals 42.
Let X be the number.
(3 + X) times 7 = 42
7(3 + X) = 42

3. A
2 + a number divided by 7.
(2 + X) divided by 7.
(2 + X)/7

4. B
Six times a number plus five is the same as saying six times (a number plus five). Or,
6 * (a number plus five). Let X be the number so,
6(X + 5).

5. A
The ratio between black and blue pens is 7 to 28 or 7:28. Bring to the lowest terms by dividing both sides by 7 gives 1:4.

6. A
At 100% efficiency 1 machine produces 1450/10 = 145 m of cloth.

At 95% efficiency, 4 machines produce 4 * 145 * 95/100 = 551 m of cloth.

At 90% efficiency, 6 machines produce 6 * 145 * 90/100 = 783 m of cloth.

Total cloth produced by all 10 machines = 551 + 783 = 1334 m

Since the information provided and the question are based on 8 hours, we did not need to use time to reach the answer.

7. D
https://youtu.be/Es3Yg5pfYeY

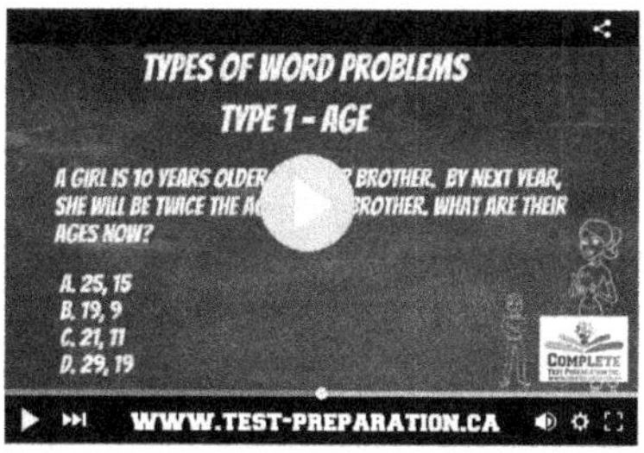

To find the total turnout in all three polling stations, we need to proportion the number of voters to the number of all registered voters.

Number of total voters = 945 + 860 + 1210 = 3015

Number of total registered voters = 1270 + 1050 + 1440 = 3760

Percentage turnout over all three polling stations = 3015 * 100/3760 = 80.19%

Checking the answers, we round 80.19 to the nearest whole number: 80%

8. D
https://youtu.be/syDAMxmkYgY

This is a simple direct proportion problem:
If Lynn can type 1 page in p minutes,

she can type x pages in 5 minutes

We do cross multiplication: x * p = 5 * 1

Then,

x = 5/p

9. A
This is an inverse ratio problem.

1/x = 1/a + 1/b where a is the time Sally can paint a house, b is the time John can paint a house, x is the time Sally and John can together paint a house.

So,

1/x = 1/4 + 1/6 … We use the least common multiple in the denominator that is 24:

1/x = 6/24 + 4/24

1/x = 10/24

x = 24/10

x = 2.4 hours.

In other words; 2 hours + 0.4 hours = 2 hours + 0.4•60 minutes

= 2 hours 24 minutes

10. D
https://youtu.be/I_FaJPJzepE

The cost of the dishwasher = $450

15% discount amount = 450•15/100 = $67.5

The discounted price = 450 – 67.5 = $382.5

20% additional discount amount on lowest price = 382.5•20/100 = $76.5

So, the final discounted price = 382.5 - 76.5 = $306.00

11. D
Original price = x,
80/100 = 12590/X,
80X = 1259000,
X = 15,737.50.

12. D
https://youtu.be/EF92e6V4mAA

We are given that each of the n employees earns s amount of salary weekly.

This means that one employee earns s salary weekly. So; Richard has 'ns' amount of money to employ n employees for a week.

We are asked to find the number of days n employees can be employed with x amount of money. We can do simple direct proportion:

If Richard can employ n employees for 7 days with 'ns' amount of money,

Richard can employ n employees for y days with x amount of money … y is the number of days we need to find.

We can cross multiply:

$y = (x * 7)/(ns)$

$y = 7x/ns$

13. B
The distribution is done at three different rates and in three different amounts:

\$6.4 per 20 kilograms to 15 shops … 20•15 = 300 kilograms distributed

\$3.4 per 10 kilograms to 12 shops … 10•12 = 120 kilograms distributed

550 - (300 + 120) = 550 - 420 = 130 kilograms left. This amount is distributed in 5 kilogram portions. So, this means that there are 130/5 = 26 shops.

\$1.8 per 130 kilograms.

We need to find the amount he earned overall these distributions.

\$6.4 per 20 kilograms : 6.4 * 15 = \$96 for 300 kilograms

\$3.4 per 10 kilograms : 3.4 *12 = \$40.8 for 120 kilograms

\$1.8 per 5 kilograms : 1.8 * 26 = \$46.8 for 130 kilograms

So, he earned 96 + 40.8 + 46.8 = \$ 183.6

The total distribution cost is given as \$10

The profit is found by: Money earned - money spent ... It is important to remember that he bought 550 kilograms of potatoes for \$165 at the beginning:

Profit = 183.6 - 10 - 165 = \$8.6

14. B
We check the fractions taking place in the question. We see that there is a "half" (that is 1/2) and 3/7. So, we multiply the denominators of these fractions to decide how to name the total money. We say that Mr. Johnson has 14x at the beginning; he gives half of this, meaning 7x, to his family. \$250 to his landlord. He has 3/7 of his money left. 3/7 of 14x is equal to:

14x * (3/7) = 6x

So,

Spent money is: 7x + 250

Unspent money is: 6x

Total money is: 14x

Write an equation: total money = spent money + unspent money

14x = 7x + 250 + 6x

14x - 7x - 6x = 250

x = 250

We are asked to find the total money that is 14x:

14x = 14 * 250 = $3500

15. D
First calculate total square feet, which is 15•24 = 360 ft2. Next, convert this value to square yards, (1 yards2 = 9 ft2) which is 360/9 = 40 yards2. At $0.50 per square yard, the total cost is 40 * 0.50 = $20.

16. D
Price increased by $5 ($25-$20). To calculate the percent increase:
5/20 = X/100
500 = 20X
X = 500/20
X = 25%

17. D
2 glasses are broken for 43 customers so 1 glass breaks for every 43/2 customers served, therefore 10 glasses implies (43/2) * 10 = 215 customers.

18. D
As the lawn is square, the length of one side will be the square root of the area. √62,500 = 250 meters. So, the perimeter is found by 4 times the length of the side of the square:

250 * 4 = 1000 meters.

Since each meter costs $5.5, the total cost of the fence will be 1000 * 5.5 = $5,500.

19. D
The question asks for approximate cost, so work with round numbers. The jacket costs $545.00 so we can round up to $550. 10% of $550 is 55. We can round down to $50, which is easier to work with. $550 - $50 is

$500. The jacket will cost about $500.

The actual cost will be 10% X 545 = $54.50
545 – 54.50 = $490.50

20. D
Let us denote Sarah's weight by "x." Then, since she weighs 25 pounds more than Tony, so he will be x - 25. They together weigh 205 pounds which means that the sum of the two representations will be equal to 205:

Sarah : x

Tony : x - 25

x + (x - 25) = 205 … by arranging this equation we have:

x + x - 25 = 205

2x - 25 = 205 … we add 25 to each side to have x term alone:

2x - 25 + 25 = 205 + 25

2x = 230

x = 230/2

x = 115 pounds → Sarah weighs 115 pounds. Since 1 pound is 0.4535 kilograms, we need to multiply 115 by 0.4535 to have her weight in kilograms:

x = 115 * 0.4535 = 52.1525 kilograms → this is equal to 52 when rounded to the nearest whole number.

21. B
(Amount Spent) $420 + $3000 (Balance) = $3,420.00

22. C

Total weight of 13 students with average 42 will be = 42 * 13 = 546 kg.

The total weight of the remaining 2 will be found by subtracting the total weight of 13 students from the total weight of 15 students: 640.5 - 546 = 94.5 kg.

94.5 = the total weight of two students. One of these students weigh 48 kg, so;

The weight of the other will be = 94.5 – 48 = 46.5 kg

23. C

Total expense is $2000 and we are informed that $5 is spent per meter. Combining these two information, we know that the total length of the fence is 2000/5 = 400 meters.

The fence is built around a square field. If one side of the square is “a,” the perimeter of the square is “4a.” Here, the perimeter is equal to 400 meters. So,

$400 = 4a$

$100 = a \rightarrow$ this means that one side of the square is equal to 100 meters

24. B

Let the number of oranges in the basket before additions = x

Then: X + 8x/5 = 130

5x + 8x = 650

650 = 13x

X = 50

25. D

The income tax per year is $9,000. So, the income tax per month is 9,000/12 = $750.

This person earns $25,000 per month and pays $750 income tax. We need to find the rate of the income tax:

Tax rate: 750 * 100/25,000 = 3%
Government increased this rate by 0.5% so it became 3.5%.

The income of the person per month is increased $11,000 so it became:

$25,000 + $11,000 = $36,000.

The new monthly income tax is: 36,000 * 3.5/100 = $1260.

Amount of increase in tax per month is:
$1260 - $750 = $510.

X = 50

25. D
The income tax per year is $9,000. So, the income tax per month is 9,000/12 = $750.

This person earns $25,000 per month and pays $750 income tax. We need to find the rate of the income tax:

Tax rate: 750 * 100/25,000 = 3%
Government increased this rate by 0.5% so it became 3.5%.

The income of the person per month is increased $11,000 so it became:

$25,000 + $11,000 = $36,000.

The new monthly income tax is: 36,000 * 3.5/100 = $1260.

Amount of increase in tax per month is:
$1260 - $750 = $510.

Basic Geometry

THE BASIC GEOMETRY COVERS THE FOLLOWING:

- Slope of a line
- Identify linear equations from a graph
- Calculate perimeter, circumference and volume
- Solve problems using the Pythagorean theorem
- Solve real world problems using the properties of geometric shapes
- Similarity and congruence

Cartesian Plane, Coordinate Grid and Plane

To locate points and draw lines and curves, we use the coordinate plane. It also called Cartesian coordinate plane. It is a two-dimensional surface with a coordinate grid in it, which helps us to count the units. For the counting of those units, we use x-axis (horizontal scale) and y-axis (vertical scale).

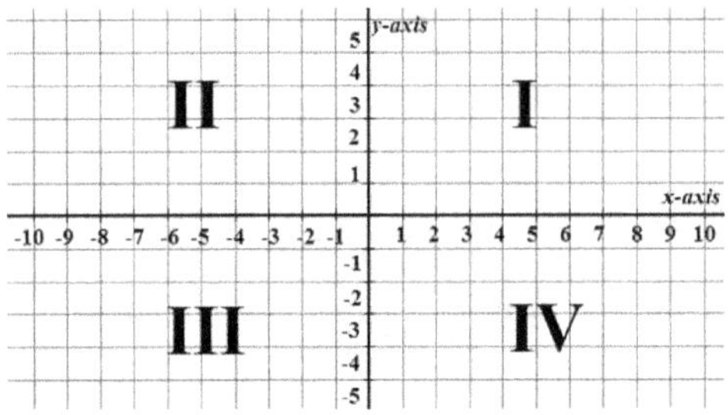

The whole system is called a coordinate system which is divided into 4 parts, called quadrants. The quadrant where all numbers are positive is the 1st quadrant (I), and if we go counterclockwise, we mark all 4 quadrants.

The location of a dot in the coordinate system is represented by coordinates. Coordinates are represented as a pair of numbers, where the 1st number is located on the x-axis and the 2nd number is located on the y-axis. So, if a dot A has coordinates a and b, then we write:

A=(a,b) or A(a,b)

The point where x-axis and y-axis intersect is called an origin. The origin is the point from which we measure the distance along the x and y axes.

In the Cartesian coordinate system we can calculate the distance between 2 given points. If we have dots with coordinates:
A=(a,b)
B=(c,d)

Then the distance d between A and B can be calculated by the following formula:

$$d=\sqrt{(c-a)^2+(d-b)^2}$$

Cartesian coordinate system is used for the drawing of 2-dimentional shapes, and is also commonly used for functions.

Example:

Draw the function y = (1 - x)/2

To draw a linear function, we need at least 2 points.
If we put that x=0 then value for y would be:

$$y=\frac{1-x}{2}=\frac{1-0}{2}=\frac{1}{2}$$

We found the 1st point, let's name it A, with following coordinates:

A = (0,1/2)

To find the 2nd point, we can put that x=1. Here, the value for y would be:

$$y=\frac{1-x}{2}=\frac{1-1}{2}=\frac{0}{2}=0$$

If we denote the 2nd point with B, then the coordinates for this point are:

B=(1,0)

Since we have 2 points necessary for the function, we find them in the coordinate system and we connect them with a line that represents the function,

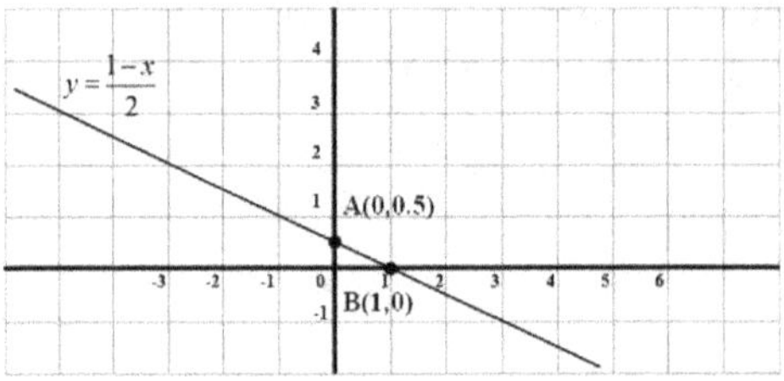

Perimeter Area and Volume

Perimeter and Area (2-dimentional shapes)

Perimeter of a shape determines the length around that shape, while the area includes the space inside the shape.

Rectangle:

$P = 2a + 2b$

$A = ab$

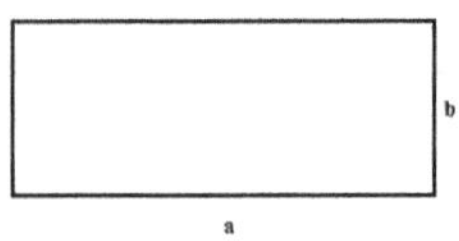

Square

$P = 4a$

$A = a^2$

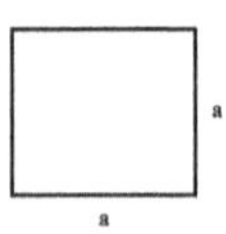

Parallelogram

$$P = 2a + 2b$$
$$A = ah_a = bh_b$$

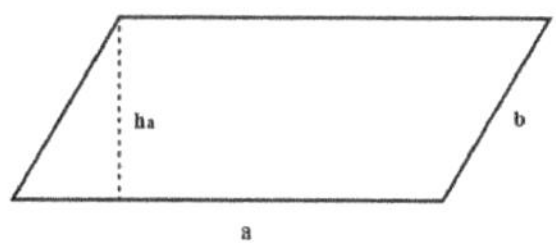

Rhombus

$$P = 4a$$
$$A = ah = \frac{d_1 d_2}{2}$$

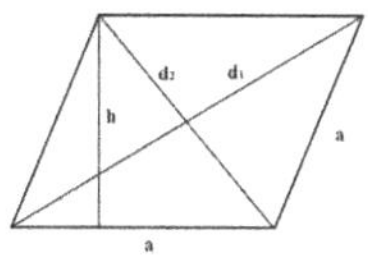

Triangle

$$P = a + b + c$$
$$A = \frac{ah_a}{2} = \frac{bh_b}{2} = \frac{ch_c}{2}$$

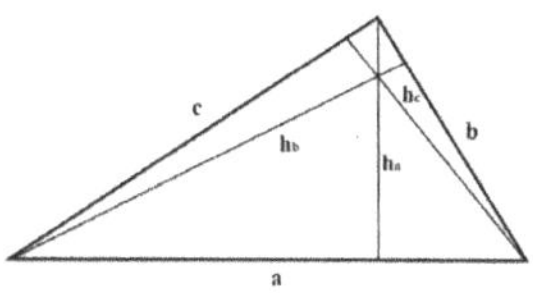

Equilateral Triangle

$$P = 3a$$
$$A = \frac{a^2\sqrt{3}}{4}$$

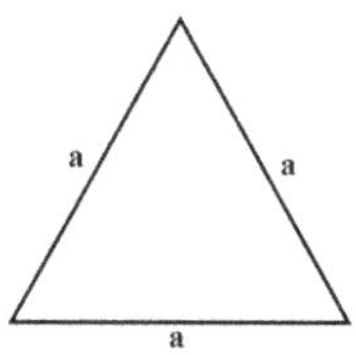

Trapezoid

$$P = a + b + c + d$$
$$A = \frac{a+b}{2}h$$

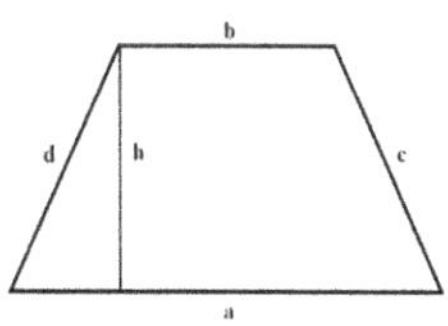

Circle

$P = 2r\pi$

$A = r^2\pi$

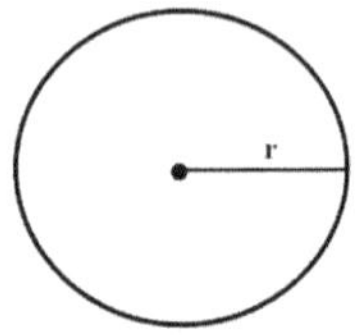

Area and Volume (3-dimentional shapes)

To calculate the area of a 3-dimentional shape, we calculate the areas of all sides and then we add them all.

To find the volume of a 3-dimentional shape, we multiply the area of the base (B) and the height (H) of the 3-dimentional shape.

$$V = BH$$

In case of a pyramid and a cone, the volume would be divided by 3.

$$V = BH/3$$

Here are some of the 3-dimentional shapes with formulas for their area and volume:

Cuboids

$A = 2(ab + bc + ac)$

$V = abc$

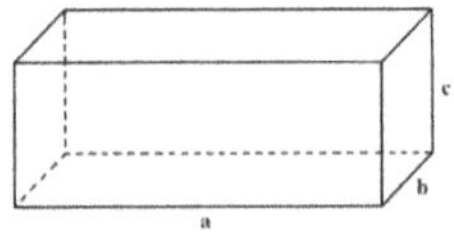

Cube

$A = 6a^2$

$V = a^3$

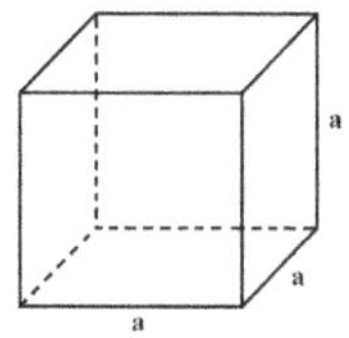

Pyramid

$A = ab + ah_a + bh_b$

$V = \frac{abH}{3}$

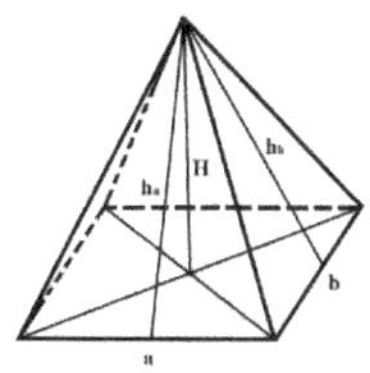

Cylinder

$A = 2r^2\pi + 2r\pi H$

$V = r^2\pi H$

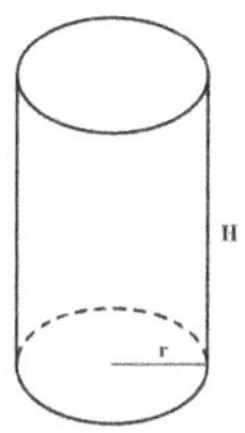

Cone

$A = (r + s)r\pi$

$V = \frac{r^2\pi H}{3}$

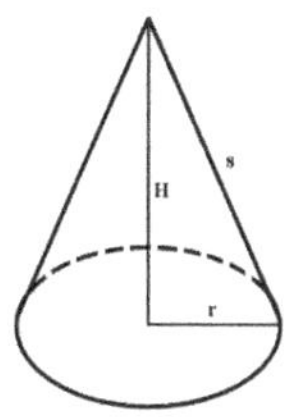

https://youtu.be/ZWZ4NoCH7s8

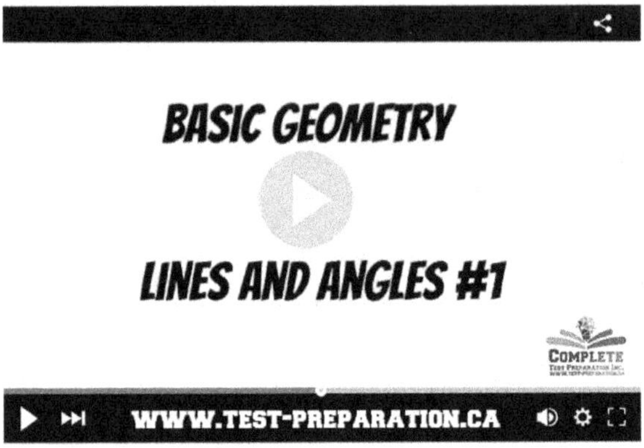

https://youtu.be/UNPse0R9Qto

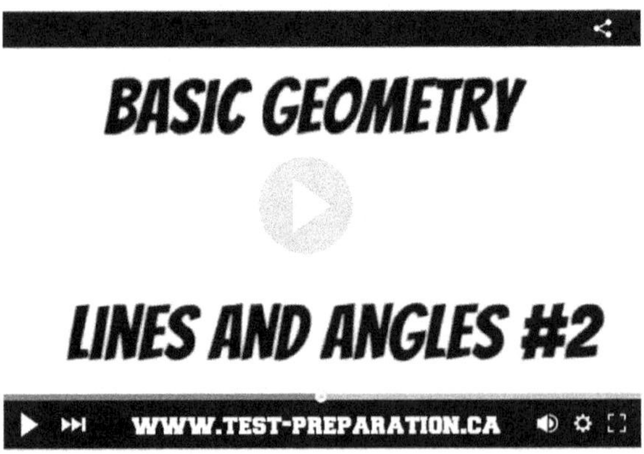

Pythagorean Geometry

If we have a right triangle ABC, where its sides (legs) are a and b and c is a hypotenuse (the side opposite the right angle), then we can establish a relationship between these sides using the following formula:

$$c^2 = a^2 + b^2$$

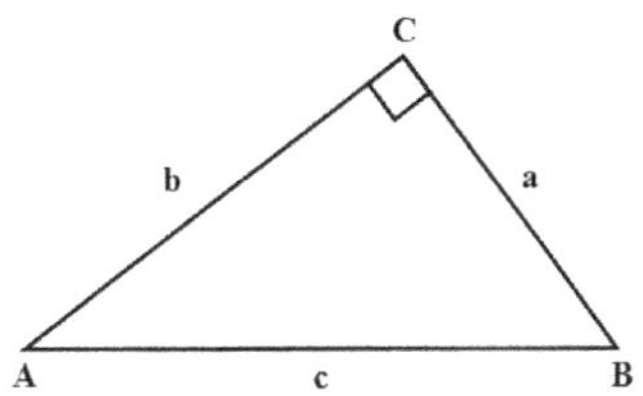

This formula is proven in the Pythagorean Theorem. There are many proofs of this theorem, but we'll look at just one geometrical proof:

If we draw squares on the right triangle's sides, then the area of the square upon the hypotenuse is equal to the sum of the areas of the squares that are upon other two sides of the triangle. Since the areas of these squares are a^2, b^2 and c^2, that is how we got the formula above.

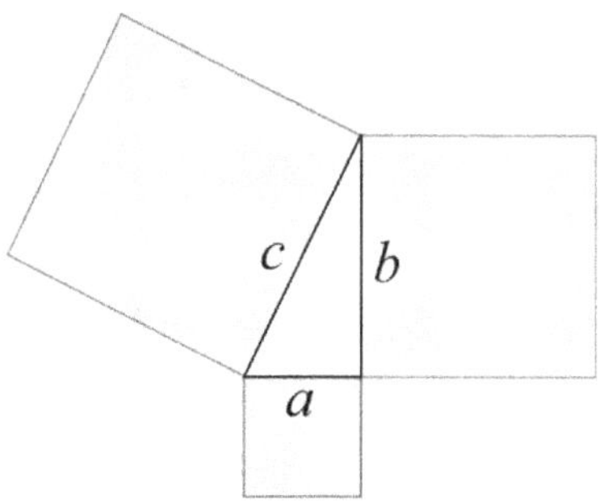

One of the famous right triangles is one with sides 3, 4 and 5. And we can see here that:

$3^2 + 4^2 = 5^2$
$9 + 16 = 25$
$25 = 25$

Example Problem:

The isosceles triangle ABC has a perimeter of 18 centimeters, and the difference between its base and legs is 3 centimeters. Find the height of this triangle.

We write the information we have about triangle ABC and we draw a picture for a better understanding of the rela-

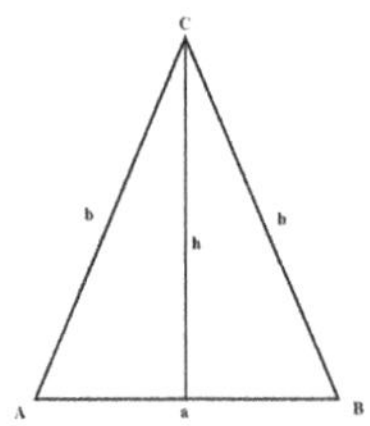

tion between its elements:

P=18 cm
a - b = 3 cm
h=?

We use the formula for the perimeter of the isosceles triangle, since that is what is given to us:
P = a + 2b = 18 cm

Notice that we have 2 equations with 2 variables, so we can solve it as a system of equations:

a + 2b = 18
a – b = 3 / a + 2b = 18
2a - 2b = 6 / a + 2b + 2a - 2b = 18 + 6

$3a = 24$
$a = 24/3 = 8$ cm

Now we go back to find b:
$a - b = 3$
$8 - b = 3$
$b = 8 - 3$
$b = 5$ cm

Using Pythagorean Theorem, we can find the height using a and b, because the height falls on the side a at the right angle. Notice that height cuts side a exactly in half, and that's why we use in the formula a/2. Here, b is our hypotenuse, so we have:

$b^2 = (a/2)^2 + h^2$
$h^2 = b^2 - (a/2)^2$
$h^2 = 5^2 - (8/2)^2$
$h^2 = 5^2 - (8/2)^2$
$h^2 = 25 - 4^2$
$h^2 = 26 - 16$
$h^2 = 9$
$h = 3$ cm.

Quadrilaterals

Quadrilaterals are 2-dimentional geometrical shapes that have 4 sides and 4 angles. There are many types of quadrilaterals, depending on the length of its sides, if they are parallel, and the size of its angles. All quadrilaterals have the following properties:

Sum of all interior angles is 360^0

Sum of all exterior angles is 360^0

A quadrilateral is a parallelogram is it fulfills at least one of the following conditions:

- Angles on each side are supplementary
- Opposite angles are equal
- Opposite sides are equal
- Diagonals intersect each other exactly in half

Here are some of the quadrilaterals:

Square

All sides are equal
All angles are right angles

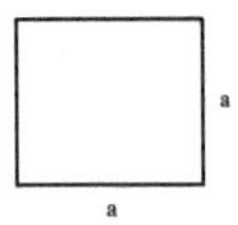

Rectangle

2 pairs of equal sides
All angles are right angles

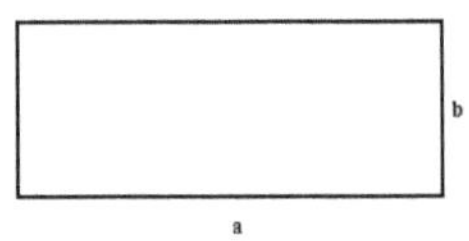

Parallelogram

2 pairs of equal sides
Opposite angles are equal

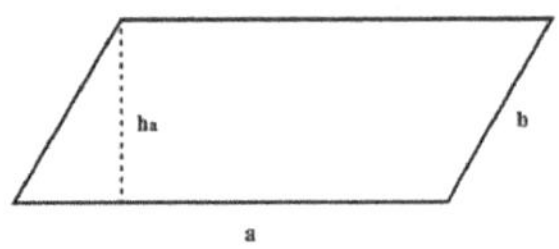

Rhombus

All sides are equal
Opposite angles are equal

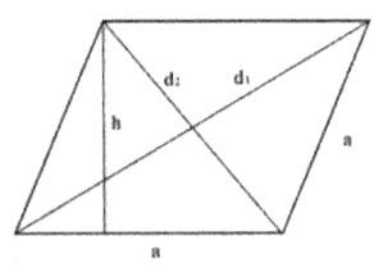

Trapezoid

One pair of parallel sides

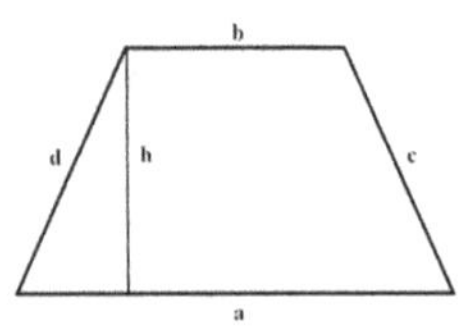

Example Problem
Find all angles of a parallelogram if one angle is greater than the other one by 40^0.

First, we draw an image of a parallelogram:

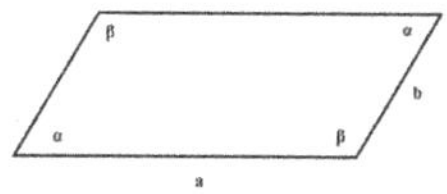

We denote angles by α and β, Since this is a parallelogram, the opposite angles are equal.

We are given that one angle is greater than the other one by 40^0, so we can write:

$\beta = \alpha + 40^0$

We solve this problem in two ways:
1) The sum of all internal angles of every quadrilateral is 360^0. There are 2 α and 2 β. So we have:
$2\alpha + 2\beta = 360^0$
Now, instead of β we write $\alpha + 40$:
$2\alpha + 2(\alpha + 40^0) = 360^0$
$2\alpha + 2\alpha + 80^0 = 360^0$
$4\alpha = 360^0 - 80^0$
$4\alpha = 280^0$
$\alpha = 280^0 / 4$
$\alpha = 70^0$
Now we can find β from α:
$\beta = \alpha + 40^0$
$\beta = 70^0 + 40^0$
$\beta = 110^0$

2) One condition for parallelogram is " Angles on each side are supplementary" and we can use that to find these angles:

$\alpha + \beta = 180^0$
$\alpha + \alpha + 40^0 = 180^0$
$2\alpha = 180^0 - 40^0$
$2\alpha = 140^0$
$\alpha = 70^0$

Now we find β:
$\beta = \alpha + 40^0$
$\beta = 70^0 + 40^0$
$\beta = 110^0$

Congruence

Two geometrical shapes are congruent if their elements (sides and angles) are equal, but they don't have to have the same direction. So, here we are only interested in the size and shapes. For example, angles α and β are congruent if they have the same size, but not necessarily the same direction:

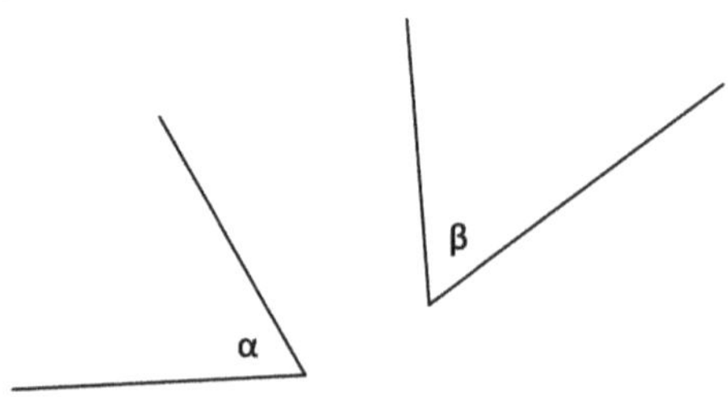

We write the congruence like this: $a \cong \beta$ $\alpha \cong \beta$
We can also say that 2 triangles are congruent if their appropriate elements are equal:

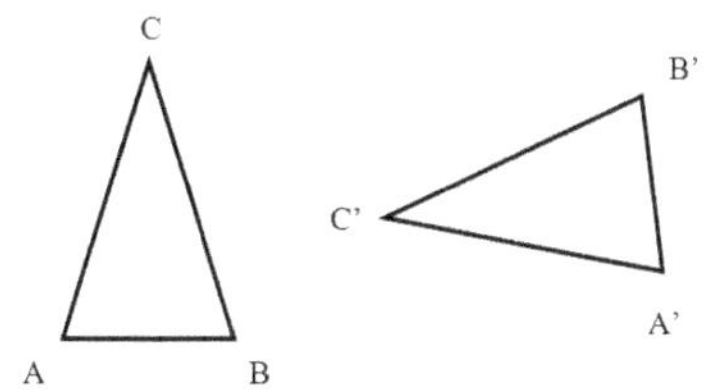

So, we write: $ABC \cong A'B'C'$

The most important congruence rules are concerning triangles, and they are used for congruence of almost all 2D shapes (quadrilaterals, hexagons etc). There are 4 rules:

1. Side-Angle-Side (SAS)
If 2 triangles have 2 equal sides and an equal angles that are between those sides, then we can conclude that these 2 triangles are congruent.

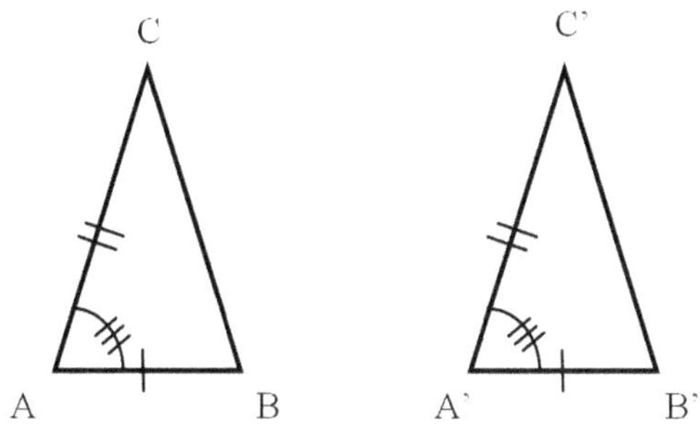

2. Side-Side-Side (SSS)
If 2 triangles have all 3 sides that are equal, then these 2 triangles are congruent.

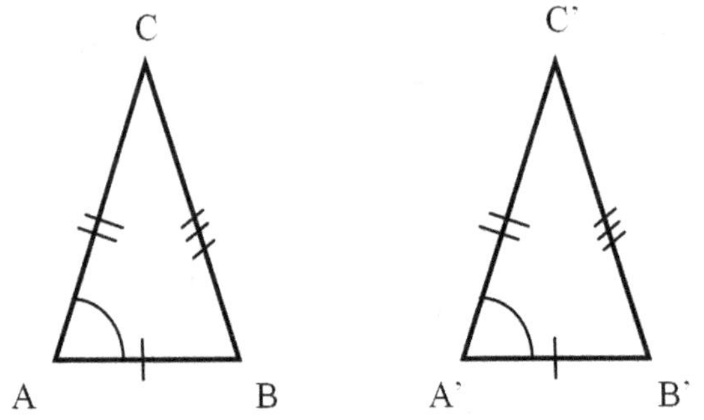

3. Angle-Side-Angle (ASA)

If 2 triangles have 2 equal angles and an equal side that is between them, then these 2 triangles are congruent.

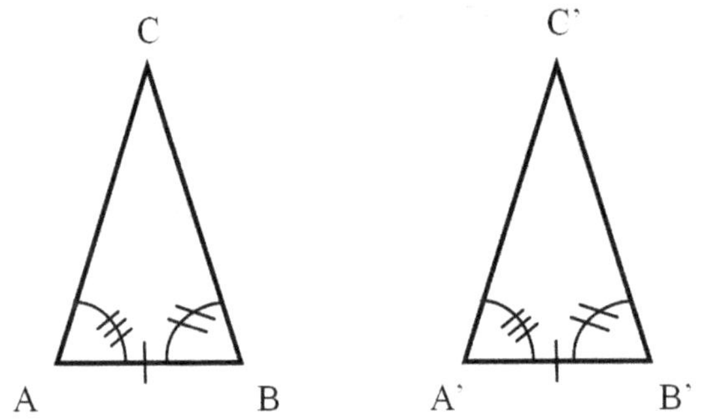

4. Side-Side-Angle (SSA)

If 2 triangles have 2 equal sides, and an equal angle that is not between those 2 sides, then these 2 triangles are congruent.

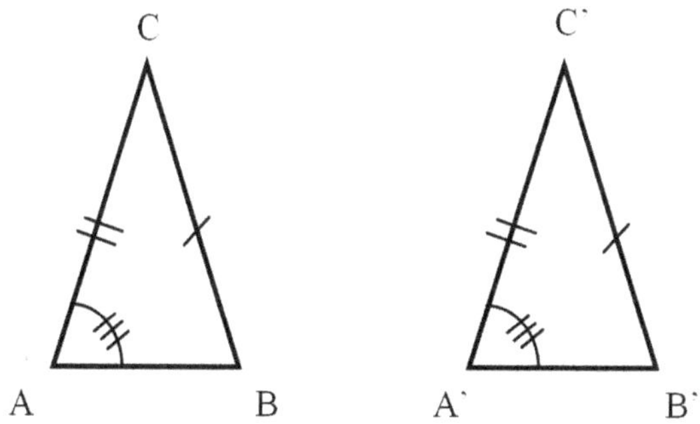

Let's look at an example question.

Consider the 2 triangles ABC and A'B'C'.
We see that we have a case of SAS:

So we can conclude that triangles ABC and A'B'C' are congruent. So, we write:

$ABC \cong A'B'C'$

Fundamental Theorems in Geometry

Pythagorean Theorem: In a right triangle, the square of the length of the hypotenuse equals the sum of the squares of the other two sides.
Triangle Sum Theorem: The sum of the interior angles of a triangle is 180 degrees.
Congruent Triangles (SSS, SAS, ASA, AAS, HL): Two triangles are congruent if their corresponding sides and/or angles are equal according to specific criteria (Side-Side-Side, Side-Angle-Side, Angle-Side-Angle, Angle-Angle-Side, Hypotenuse-Leg for right triangles).
Isosceles Triangle Theorem: If two sides of a triangle are congruent, then the angles opposite those sides are congruent.
Alternate Interior Angles Theorem: If two parallel lines are cut by a transversal, alternate interior angles are equal.
Corresponding Angles: If two parallel lines are cut by a transversal, corresponding angles are equal.
Exterior Angle Theorem An exterior angle of a triangle is equal to the sum of the two non-adjacent interior angles.
Midpoint Theorem: The line segment joining the midpoints of two sides of a triangle is parallel to the third side and half as long.
Angle Bisector Theorem: The angle bisector of a triangle divides the opposite side into two segments that are proportional to the adjacent sides.
Similar Triangles Theorem: If two triangles have corresponding angles equal, their sides are proportional, and the triangles are similar.
Parallel Postulate: Through a point not on a given line, there is exactly one line parallel to the given line.

Circle Theorems:

b. An angle inscribed in a semicircle is a right angle.

c. The measure of a central angle is equal to the measure of the arc it intercepts.

d. Angles in the same segment of a circle are equal.

Important Equations in Geometry

Area of a Rectangle: A = length × width
Area of a Triangle: A = (base × height) ÷ 2
Area of a Circle: A = π × $radius^2$
Circumference of a Circle: C = 2 × π × radius
Area of a Parallelogram: A = base × height
Area of a Trapezoid: A = ($base_1$ + $base_2$) × height ÷ 2
Pythagorean Theorem: $a^2 + b^2 = c^2$ (for right triangles, where c is the hypotenuse)
Volume of a Rectangular Prism: V = length × width × height
Volume of a Cylinder: V = π × $radius^2$ × height
Volume of a Sphere: V = (4/3) × π × $radius^3$
Surface Area of a Sphere: SA = 4 × π × $radius^2$
Volume of a Cone: V = (1/3) × π × $radius^2$ × height
Volume of a Pyramid: V = (1/3) × base area × height
Distance Formula (between two points):
$d = \sqrt{[(x_2 - x_1)^2 + (y_2 - y_1)^2]}$
Midpoint Formula (between two points):
$M = ((x_1 + x_2)/2, (y_1 + y_2)/2)$

Answer Sheet

1.	A	B	C	D	21.	A	B	C	D
2.	A	B	C	D	22.	A	B	C	D
3.	A	B	C	D	23.	A	B	C	D
4.	A	B	C	D	24.	A	B	C	D
5.	A	B	C	D	25.	A	B	C	D
6.	A	B	C	D	26.	A	B	C	D
7.	A	B	C	D	27.	A	B	C	D
8.	A	B	C	D	28.	A	B	C	D
9.	A	B	C	D	29.	A	B	C	D
10.	A	B	C	D	30.	A	B	C	D
11.	A	B	C	D	31.	A	B	C	D
12.	A	B	C	D	32.	A	B	C	D
13.	A	B	C	D	33.	A	B	C	D
14.	A	B	C	D	34.	A	B	C	D
15.	A	B	C	D	35.	A	B	C	D
16.	A	B	C	D	36.	A	B	C	D
17.	A	B	C	D	37.	A	B	C	D
18.	A	B	C	D	38.	A	B	C	D
19.	A	B	C	D	39.	A	B	C	D
20.	A	B	C	D	40.	A	B	C	D

Geometry Practice Questions

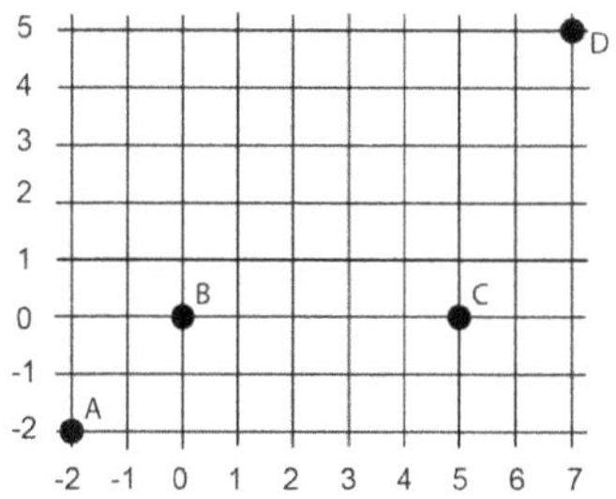

1. Which of the above points represents the origin?

a. A
b. B
c. C
d. D

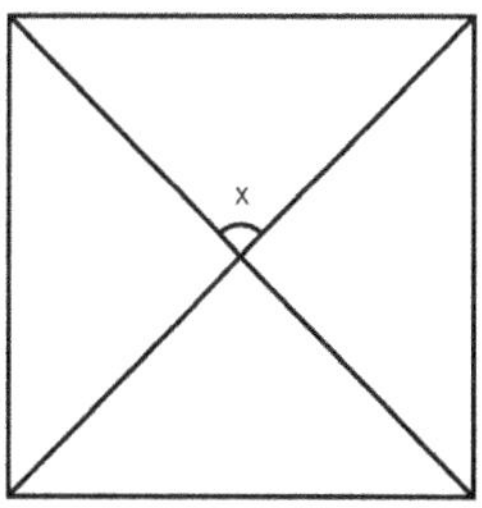

2. What is measurement of the indicated angle?

a. 45°
b. 90°
c. 60°
d. 30°

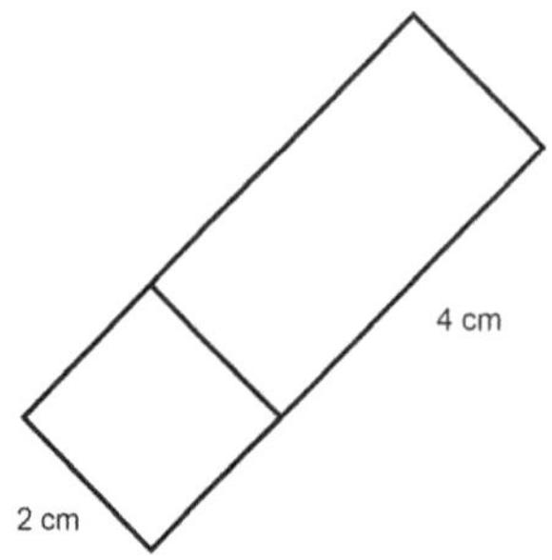

Note: figure not drawn to scale.

3. Assuming the figure with side 2 cm. is square, what is the perimeter of the above shape?

a. 12 cm
b. 16 cm
c. 6 cm
d. 20 cm

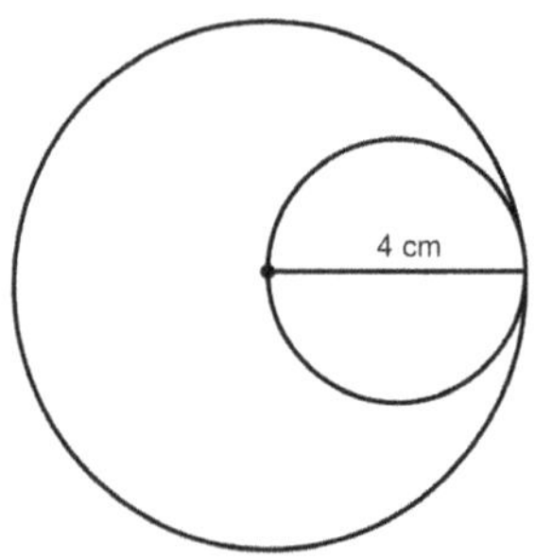

Note: Figure not drawn to scale

4. Assuming the diameter of the small circle is the radius of the larger circle, what is (area of large circle) - (area of small circle) in the figure above?

a. 8 π cm^2
b. 10 π cm^2
c. 12 π cm^2
d. 16 π cm^2

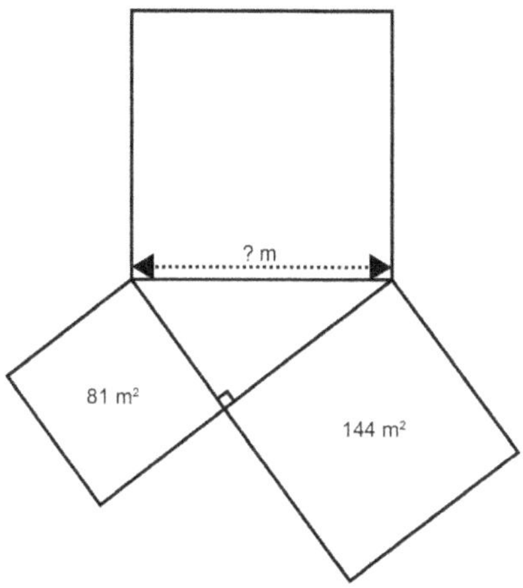

Note: Figure not drawn to scale

5. Assuming the shapes around the center right triangle are square, what is the length of each side of the indicated square above?

a. 10
b. 15
c. 20
d. 5

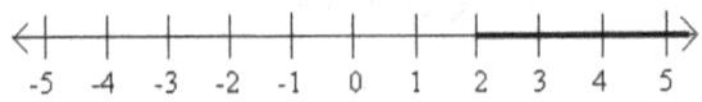

6. Choose the expression the figure represents.

a. $X \leq 1$

b. $X < 1$

c. $X > 1$

d. $X \geq 1$

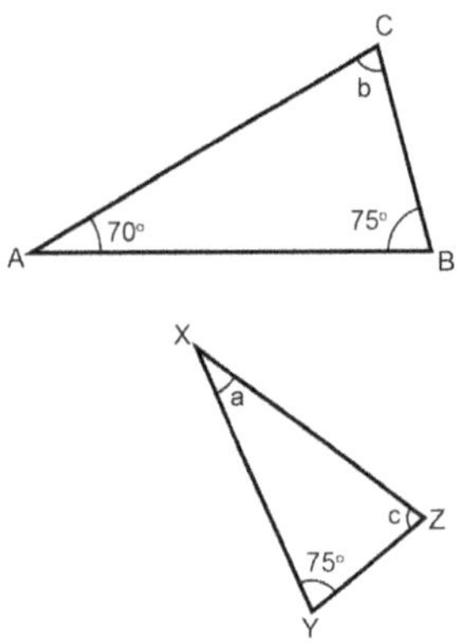

7. What are the respective values of a, b & c if both triangles are similar?

a. 70°, 70°, 35°

b. 70°, 35°, 70°

c. 35°, 35°, 35°

d. 70°, 75°, 35°

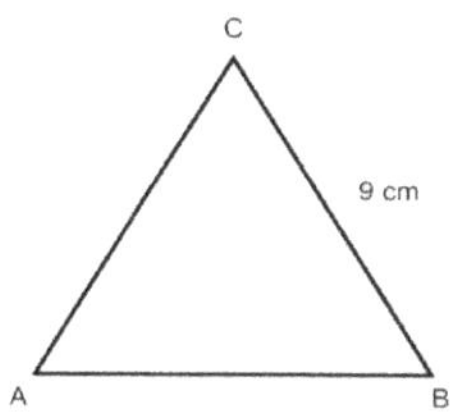

Note: figure not drawn to scale

8. What is the perimeter of the equilateral $\triangle ABC$ above?

a. 18 cm
b. 12 cm
c. 27 cm
d. 15 cm

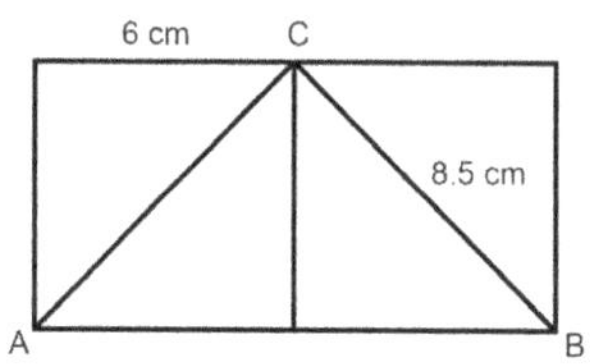

Note: figure not drawn to scale

9. Assuming the 2 quadrangles are identical rectangles, what is perimeter of $\triangle ABC$ in the above shape?

a. 25.5 cm
b. 27 cm
c. 30 cm
d. 29 cm

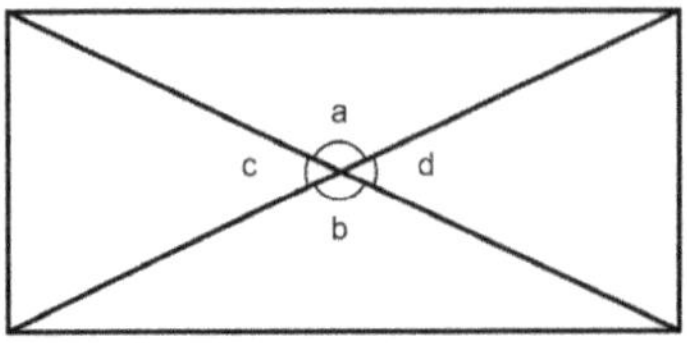

10. What is the sum of all the angles in the rectangle above?

a. 180°

b. 360°

c. 90°

d. 120°

.

Note: figure not drawn to scale

11. A tile factory makes custom tiles, shown above, from two types of stone. If a customer requires 200 tiles, how much black stone will be required?

a. $256\ m^2$

b. $2560\ m^2$

c. $2.56\ m^2$

d. $25.6\ m^2$

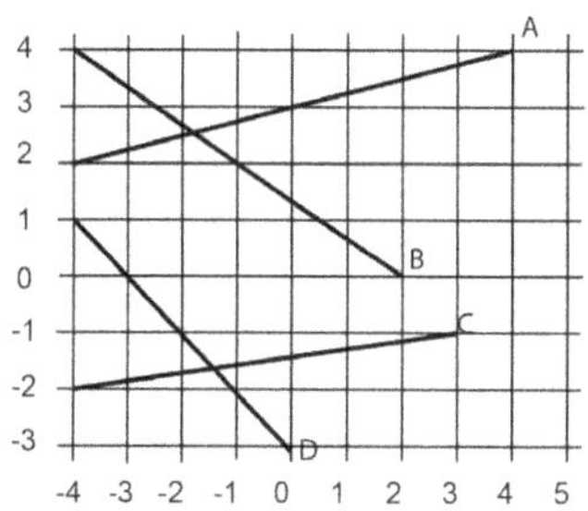

12. Which of the lines above represents the equation $2y - x = 4$?

a. A

b. B

c. C

d. D

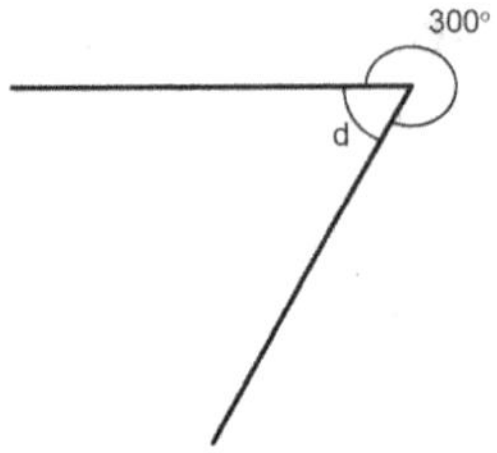

13. What is the measurement of the indicated angle?

a. 45°

b. 90°

c. 60°

d. 50°

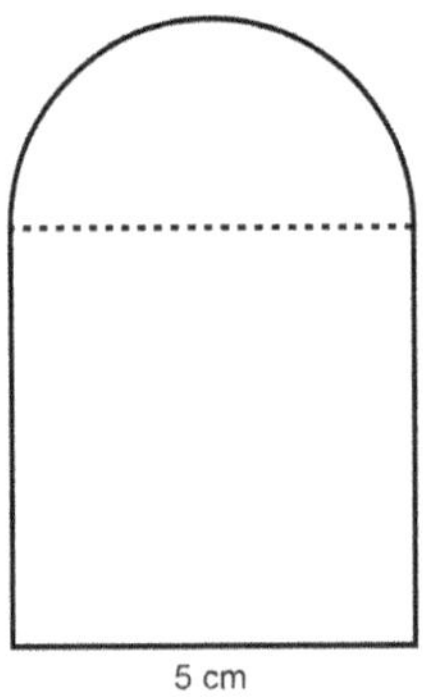

Note: figure not drawn to scale

14. What is the perimeter of the above shape, assuming the bottom portion is square?

a. 22.85 cm

b. 20 cm

c. 15 cm

d. 25.546 cm

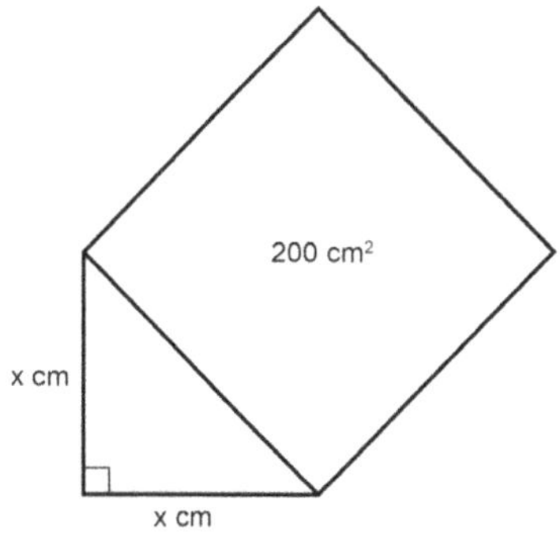

Note: Figure not drawn to scale

15. Assuming the quadrangle in the figure above is square, what is the length of the sides in the triangle above?

a. 10

b. 20

c. 100

d. 40

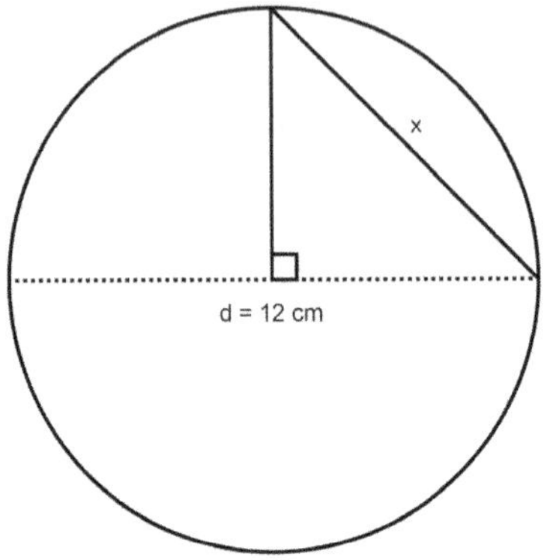

Note: Figure not drawn to scale

16. Calculate the length of side x.

a. 6.46
b. 8.48
c. 3.6
d. 6.4

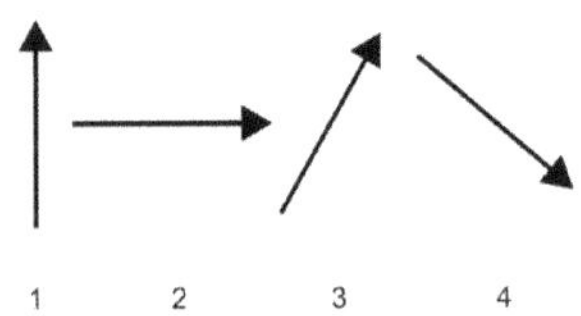

17. What is the correct order of respective slopes for the lines above?

a. Positive, undefined, negative, positive
b. Negative, zero, undefined, positive
c. Undefined, zero, positive, negative
d. Zero, positive undefined, negative

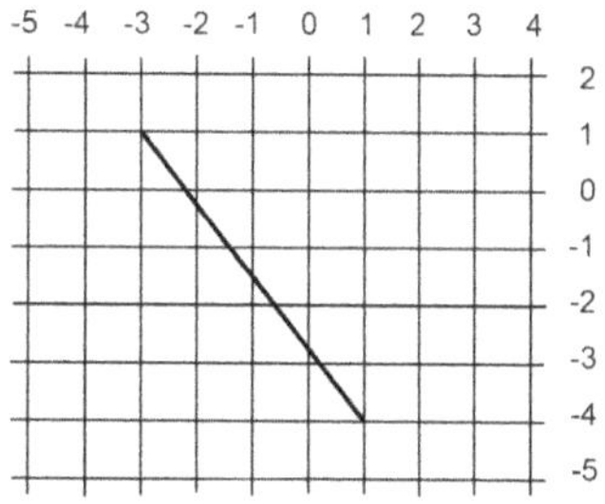

18. What is the slope of the line shown above?

a. 5/4
b. -4/5
c. -5/4
d. -4/5

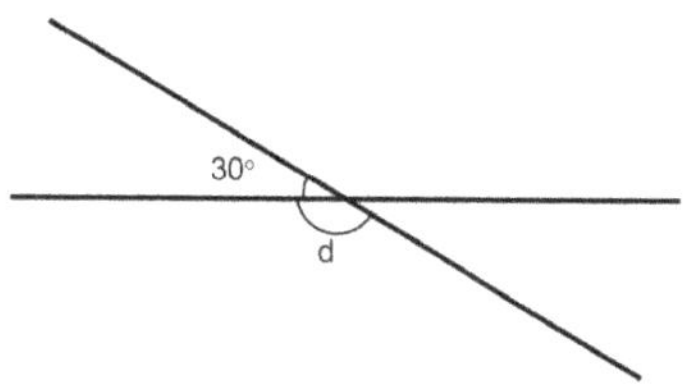

19. What is the indicated angle above?

a. 150°
b. 330°
c. 60°
d. 120°

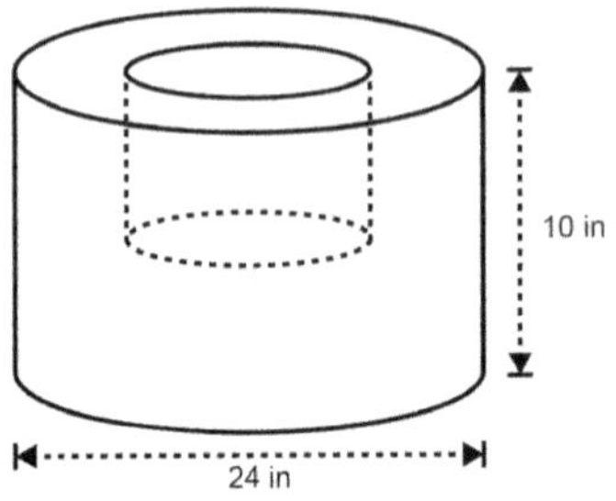

Note: figure not drawn to scale

20. What is the volume of the above solid made by a hollow cylinder that is half the size (in all dimensions) of the larger cylinder?

a. 1440 π in^3
b. 1260 π in^3
c. 1040 π in^3
d. 960 π in^3

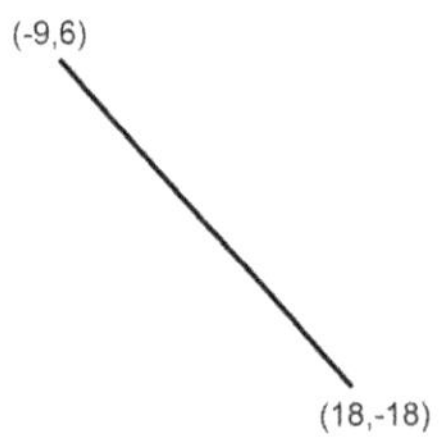

21. What is the slope of the line above?

a. -8/9
b. 9/8
c. -9/8
d. 8/9

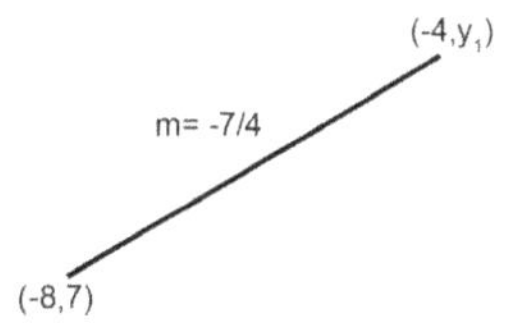

22. With the data given above, what is the value of y_1?

a. 0

b. -7

c. 7

d. 8

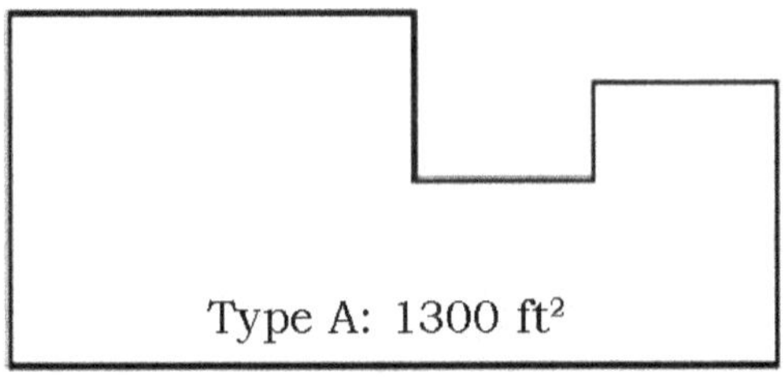

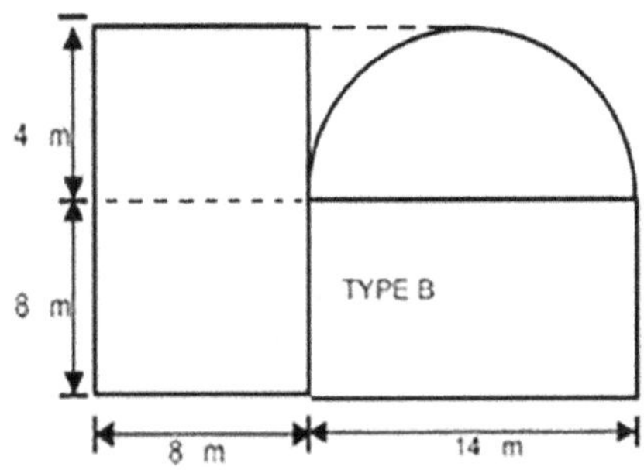

Note: Figure not drawn to scale

23. The price of houses in a certain subdivision is based on the total area. Susan is watching her budget and wants to choose the house with the lowest area. Which house type, A (1300 ft2) or B, should she choose if she would like the house with the lowest price? (1 m^2 = 10.76 ft^2 & $\pi = 22/7$)

a. Type B is smaller at 140 ft^2
b. Type A is smaller
c. Type B is smaller at 855 ft^2
d. Type B is larger

24. How much water can be stored in a cylindrical container 5 meters in diameter and 12 meters high?

Note: figure not drawn to scale

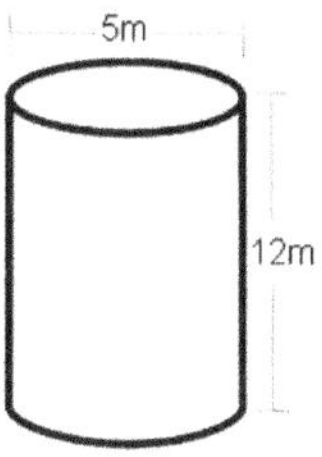

a. 235.65 m^3

b. 223.65 m^3

c. 240.65 m^3

d. 252.65 m^3

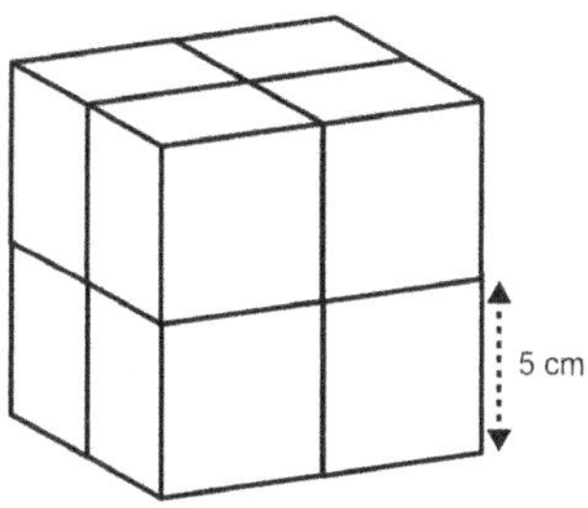

Note: figure not drawn to scale

25. Assuming the figure above is composed of cubes, what is the volume?

a. 125 cm^3

b. 875 cm^3

c. 1000 cm^3

d. 500 cm^3

26. Choose the expression the figure represents.

a. $X > 2$

b. $X \geq 2$

c. $X < 2$

d. $X \leq 2$

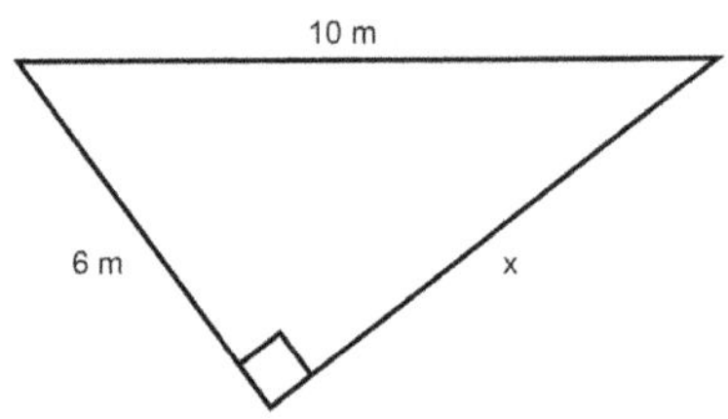

Note: figure not drawn to scale

27. What is the length of the missing side in the triangle above?

a. 6

b. 4

c. 8

d. 5

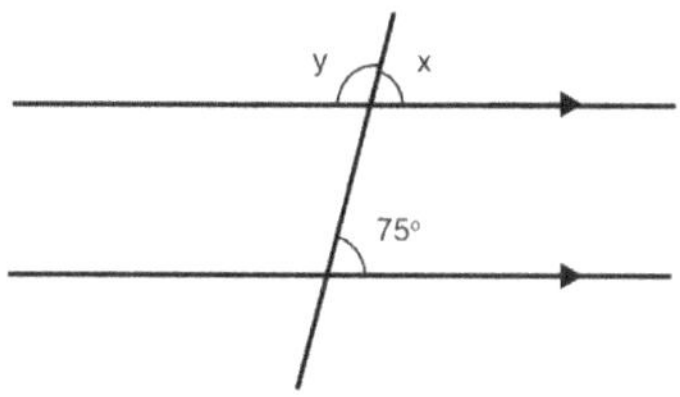

28. What is the value of the angle y?

a. 25°
b. 15°
c. 30°
d. 105°

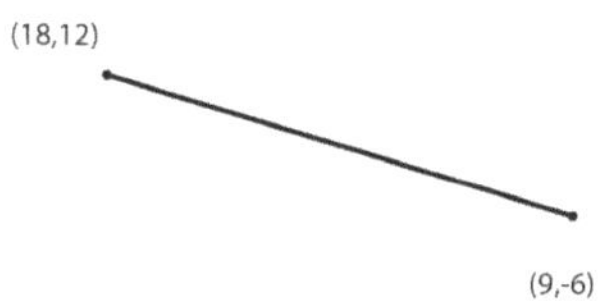

29. What is the distance between the two points?

a. ≈19
b. 22
c. ≈21
d. ≈20

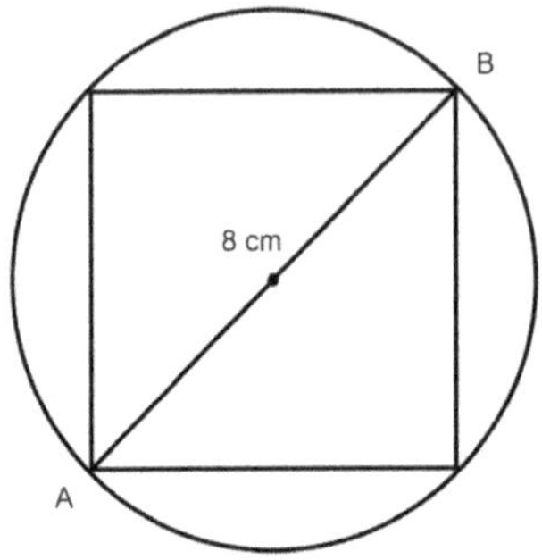

Note: figure not drawn to scale

30. What is area of the circle?

a. 4π cm²
b. 12π cm²
c. 10π cm²
d. 16π cm²

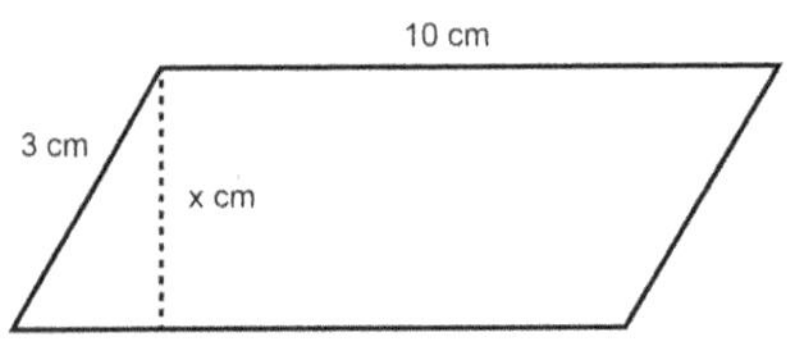

Note: figure not drawn to scale

31. What is the perimeter of the parallelogram above?

a. 12 cm
b. 26 cm
c. 13 cm
d. (13+x) cm

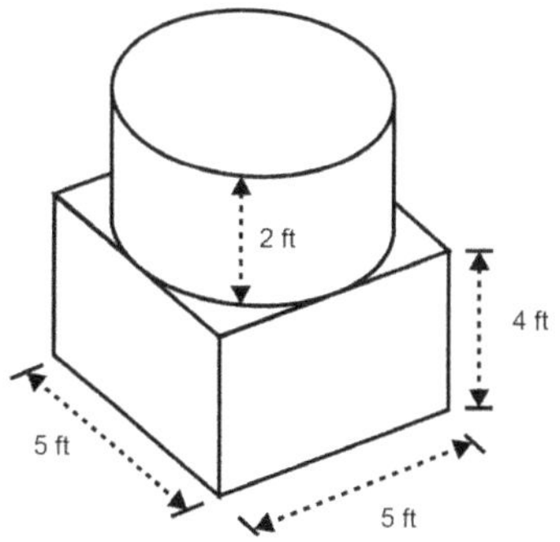

Note: figure not drawn to scale

32. What is the approximate total volume of the above solid?

a. 120 ft^3

b. 100 ft^3

c. 140 ft^3

d. 160 ft^3

33. What is the slope of the line above?

a. 1

b. 2

c. 3

d. -2

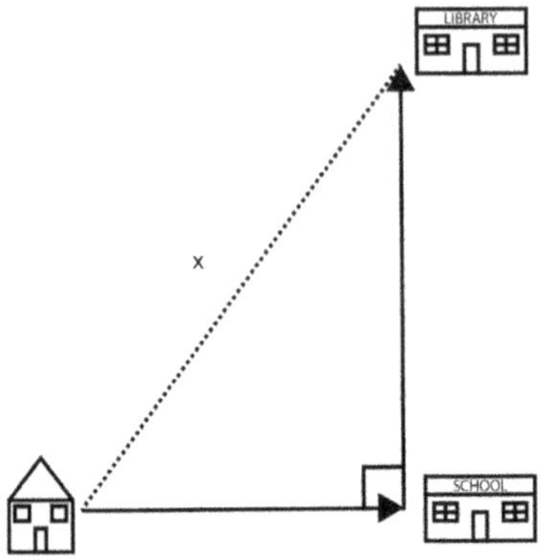

Note: figure not drawn to scale

34. Every day starting from his home Peter travels due east 3 kilometers to the school. After school he travels due north 4 kilometers to the library. What is the distance between Peter's home and the library?

a. 15 km

b. 10 km

c. 5 km

d. 12 ½ km

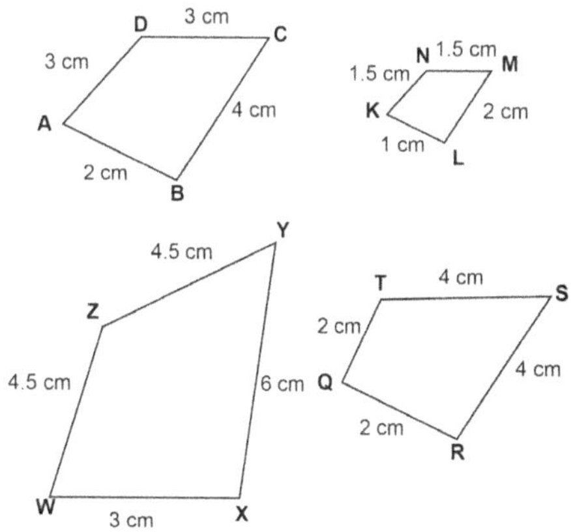

35. Which of the above quadrilaterals are similar?

a. All are similar
b. QRST, KLMN, WXYZ
c. ABCD, KLMN, WXYZ
d. None of the choices are correct

36. Consider 2 triangles, ABC and A'B'C', where:

BC = B' C'
AC = A' C'
RA = RA'

Are these 2 triangles congruent?

a. Yes
b. No
c. Not enough information

37. Consider 2 triangles ABC and A'B'C'

AB = A' B'

Angle A = Angle A'

Angle B = Angle B'

Are these 2 triangles congruent?

a. Yes

b. No

c. Not enough information

38. Rectangles ABCD and A'B'C'D' have following equal elements:

CD = C'D'
AC = A'C'

Are these 2 rectangles congruent?

39. For triangles ABC and A'B'C' we have that:

CD = C'D'
AD = A'D'

where CD is the height of the triangle ABC, and C'D' is the height of the triangle A'B'C'.

Are these 2 triangles congruent?

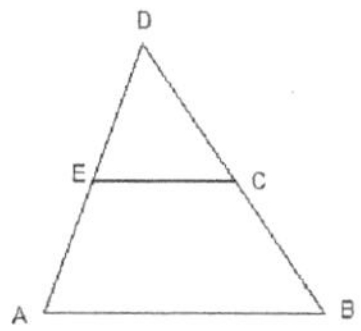

40. In the above triangle, AB // EC; |DE| = 3 cm, |EA| = 5 cm. Find the proportion of areas A(DEC) / A(ABCE).

a. 9/64

b. 9/55

c. 3/25

d. 3/5

Answer Key

1. A
Point A represents the origin.

2. A
The diagonals of a square intersect at right angles, so each angle measures 90°. Half of that angle will be 45^0

3. B
We see that there is a square with side 2 cm and a rectangle adjacent to it, with one side 2 cm (common side with the square) and the other side 4 cm. The perimeter of a shape is found by summing up all sides surrounding the shape, not adding the ones inside the shape. Three 2 cm sides from the square, and two 4 cm sides and one 2 cm side from the rectangle contribute the perimeter.

So, the perimeter of the shape is: $2 + 2 + 2 + 4 + 2 + 4 = 16$ cm.

4. C
We are given a large circle and a small circle inside it; with the diameter equal to the radius of the large one. The diameter of the small circle is 4 cm. This means that its radius is 2 cm. Since the diameter of the small circle is the radius of the large circle, the radius of the large circle is 4 cm. The area of a circle is calculated by: πr^2 where r is the radius.

Area of the small circle: $\pi(2)^2 = 4\pi$

Area of the large circle: $\pi(4)^2 = 16\pi$

The difference area is found by:

Area of the large circle - Area of the small circle = $16\pi - 4\pi = 12\pi$

5. B
We see that there are three squares forming a right triangle in the middle. Two of the squares have the areas

81 m^2 and 144 m^2. If we denote their sides a and b respectively:

$a^2 = 81$ and $b^2 = 144$. The length, which is asked, is the hypotenuse; a and b are the opposite and adjacent sides of the right angle. By using the Pythagorean Theorem, we can find the value of the asked side:

Pythagorean Theorem:

$(\text{Hypotenuse})^2 = (\text{Opposite Side})^2 + (\text{Adjacent Side})^2$

$h^2 = a^2 + b^2$

$a^2 = 81$ and $b^2 = 144$ are given. So;

$h^2 = 81 + 144$

$h^2 = 225$

$h = 15$ m

6. B
The line is pointing towards numbers less than 1. The equation is therefore, $X < 1$.

7. D
Comparing respective angles - 70°, 75°, 35°

8. C
Equilateral triangle with 9 cm. sides
Perimeter = 9 + 9 + 9 = 27 cm.

9. D
Perimeter of triangle ABC is asked.
Perimeter of a triangle = sum of all three sides.

Here, Perimeter of $\Delta ABC = |AC| + |CB| + |AB|$.

Since the triangle is located in the middle of two adjacent and identical rectangles, we find the side lengths using these rectangles:

|AB| = 6 + 6 = 12 cm

|CB| = 8.5 cm

|AC| = |CB| = 8.5 cm

Perimeter = |AC| + |CB| + |AB| = 8.5 + 8.5 + 12 = 29 cm

10. B
a + b + c + d = ?
The sum of angles around a point is 360°
a + b + c + d = 360°

11. A
Black stone for 200 tiles = 200 x [Total tile area – Inner white area(4 triangles)]

= 200 x [(16^2) - (4 x 1/2 x 8 x 8)]

= 200 x (256 - 128) = 200 x 128 = 25600 cm^2

Converting to meters – 1 cm. = 0.01 meters

= 25600/100 m^2

= 256 m^2

12. A
If a line represents an equation, all points on that line should satisfy the equation. Meaning that all (x, y) pairs present on the line should be able to verify that 2y - x is equal to 4. We can find out the correct line by trying a (x, y) point existing on each line. It is easier to choose points on the intersection of the grid lines:

Let us try the point (4, 4) on line A:

2 * 4 - 4 = 4

8 - 4 = 4

4 = 4 … this is a correct result, so the equation for line A is 2y - x = 4.

Let us try other points to check the other lines:

Point (-1, 2) on line B:

2 * 2 - (-1) = 4

4 + 1 = 4

5 = 4 … this is a wrong result, so the equation for line B is not 2y - x = 4.

Point (3, -1) on line C:

2 * (-1) - 3 = 4

-2 - 3 = 4

-5 = 4 … this is a wrong result, so the equation for line C is not 2y - x = 4.

Point (-2, -1) on line D:

2 * (-1) - (-2) = 4

-2 + 2 = 4

0 = 4 … this is a wrong result, so the equation for line D is not 2y - x = 4.

13. C

The sum of angles around a point is 360°

$d + 300 = 360^\circ$

$d = 60^\circ$

14. A

Find the perimeter of a shape made by merging a square and a semi circle. Perimeter = 3 sides of the square + 1/2 circumference of the circle.

= (3 x 5) + 1/2 (5 π)

$= 15 + 2.5\pi$

$= 15 + 7.853975$

Perimeter = 22.85 cm

15. A
If we call one side of the square "a," the area of the square will be a^2.

We know that $a^2 = 200\ cm^2$.

On the other hand; there is an isosceles right triangle. Using the **Pythagorean Theorem:**

$(\text{Hypotenuse})^2 = (\text{Adjacent Side})^2 + (\text{Opposite Side})^2$
Where the hypotenuse is equal to one side of the square. So,

$a^2 = x^2 + x^2$

$200 = 2x^2$

$200/2 = 2x^2/2$

$100 = x^2$

$x = \sqrt{100}$

$x = 10$ cm

16. B
In the question, we have a right triangle formed inside the circle. We are asked to find the length of the hypotenuse of this triangle. We can find the other two sides of the triangle by using circle properties:

The diameter of the circle is equal to 12 cm. The legs of the right triangle are the radii of the circle; so they are 6 cm long.

Using the Pythagorean Theorem:

$(\text{Hypotenuse})^2 = (\text{Adjacent Side})^2 + (\text{Opposite Side})^2$

$x^2 = r^2 + r^2$

$x^2 = 6^2 + 6^2$

$x^2 = 72$

$x = \sqrt{72}$

$x = 8.48$

17. C
Undefined, zero, positive, negative.

18. C
$$\text{Slope } (m) = \frac{\text{change in y}}{\text{change in x}}$$

$(x_1, y_1)=(-3,1)$ & $(x_2, y_2)= (1,-4)$
Slope = [-4 - 1]/[1-(-3)]= -5/4

19. A
The angles opposite both angles 30° and angle d are respectively equal to vertical angles.

$2(30° + d) = 360°$

$2d = 360° - 60°$

$2d = 300°$

$d = 150°$

20. B
Total Volume = Volume of large cylinder - Volume of small cylinder

Volume of a cylinder = area of base • height = $\pi r^2 \cdot h$

Total Volume = $(\pi * 12^2 * 10) - (\pi * 6^2 * 5) = 1440\pi - 180\pi$

$= 1260\pi$ in^3

21. A
If we know the coordinates of two points on a line, we can find the slope (m) with the below formula:

$m = (y_2 - y_1)/(x_2 - x_1)$ where (x_1, y_1) represent the coordinates of one point and (x_2, y_2) the other.

In this question:

$(-9, 6) : x_1 = -9, y_1 = 6$

$(18, -18) : x_2 = 18, y_2 = -18$

Inserting these values into the formula:

$m = (-18 - 6)/(18 - (-9)) = (-24)/(27)$ … Simplifying by 3:

$m = -8/9$

22. A
If we know the coordinates of two points on a line, we can find the slope (m) with the below formula:
$m = (y_2 - y_1)/(x_2 - x_1)$ where (x_1, y_1) represent the coordinates of one point and (x_2, y_2) the other.

In this question:

$(-4, y_1) : x_1 = -4, y_1 =$ we will find

$(-8, 7) : x_2 = -8, y_2 = 7$

$m = -7/4$

Inserting these values into the formula:

$-7/4 = (7 - y_1)/(-8 - (-4))$

$-7/4 = (7 - y_1)/(-8 + 4)$

$7/(-4) = (7 - y_1)/(-4)$ … Simplifying the denominators of both sides by -4:

$7 = 7 - y_1$

$0 = -y_1$

$y_1 = 0$

23. D
Area of Type B consists of two rectangles and a half circle. We can find these three areas and sum them up to find the total area:

Area of the left rectangle: (4 + 8) * 8 = 96 m^2

Area of the right rectangle: 14 * 8 = 112 m^2

The diameter of the circle is equal to 14 m. So, the radius is 14/2 = 7:

Area of the half circle = (1/2) * πr^2 = (1/2) * (22/7) * $(7)^2$ = (1 * 22 * 49) / (2 * 7) = 77 m^2

Area of Type B = 96 + 112 + 77 = 285 m^2

Converting this area to ft^2: 285 m^2 = 285•10.76 ft^2 = 3066.6 ft^2

Type B is (3066.6 - 1300 = 1766.6 ft^2) 1766.6 ft^2 larger than type A.

24. A
The formula of the volume of cylinder is the base area multiplied by the height. As the formula:

Volume of a cylinder = $\pi r^2 h$. Where π is 3.142, r is radius of the cross sectional area, and h is the height.

We know that the diameter is 5 meters, so the radius is 5/2 = 2.5 meters.

The volume is: V = 3.142 * 2.5^2 * 12 = 235.65 m^3.

25. C

The large cube is made up of 8 smaller cubes with 5 cm sides. The volume of a cube is found by the third power of the length of one side.

Volume of the large cube = Volume of the small cube•8

$= (5^3) * 8 = 125 * 8$

$= 1000 \text{ cm}^3$

There is another solution for this question. Find the side length of the large cube. There are two cubes rows with 5 cm length for each. So, one side of the large cube is 10 cm.

The volume of this large cube is equal to $10^3 = 1000 \text{ cm}^3$

26. A

The line is pointing towards numbers greater than 2. The equation is therefore, $X > 2$.

27. C

Pythagorean Theorem:

$(\text{Hypotenuse})^2 = (\text{Perpendicular})^2 + (\text{Base})^2$

$h^2 = a^2 + b^2$

Given: $a = 6$, $h = 10$

$h^2 = a^2 + b^2$

$b^2 = h^2 - a^2$

$b^2 = 10^2 + 6^2$

$b^2 = 100 – 36$

$b^2 = 64$

$b = 8$

28. D

Two parallel lines intersected by a third line with angles of 75°

x = 75° (corresponding angles)

x + y = 180° (supplementary angles)

y = 180° - 75°

y = 105°

29. D

The distance between two points is found by = $[(x_2 - x_1)^2+(y_2 - y_1)^2]^{1/2}$

In this question:

(18, 12) : $x_1 = 18$, $y_1 = 12$

(9, -6) : $x_2 = 9$, $y_2 = -6$

Distance= $[(9 - 18)^2 + (-6 - 12)^2]^{1/2}$

$= [(-9)^2 + (-18)^2]^{1/2}$

$= (9^2 + 2^2 \cdot 9^2)^{1/2}$

$= (9^2(1 + 4))^{1/2}$ ... We can take 9 out of the square root:

$= 9 * 5^{1/2}$

$= 9\sqrt{5}$

= 9 * 2.45

= 20.12

The distance is approximately 22 units.

30. D

We have a circle given with diameter 8 cm and a square located within the circle. We are asked to find the area of the circle for which we only need to know the length of the radius that is the half of the diameter.

Area of circle = πr^2 … r = 8/2 = 4 cm

Area of circle = $\pi * 4^2$

= 16π cm^2 … As we notice, the inner square has no role in this question.

31. B

Perimeter of a parallelogram is the sum of the sides.
Perimeter = 2(l + b)
Perimeter = 2(3 +10), 2 x 13
Perimeter = 26 cm.

32. C

Volume of a cylinder is $\pi \times r^2 \times h$

Diameter = 5 ft. so radius is 2.5 ft.

Volume of cylinder= $\pi \times 2.5^2 \times 2$

= $\pi \times 6.25 \times 2 = 12.5\ \pi$

Approximate π to 3.142

Volume of the cylinder = 39.25

Volume of a rectangle = height X width X length.
= 5 X 5 X 4 = 100

Total volume = Volume of rectangular solid + volume of cylinder

Total volume = 100 + 39.25

Total volume = 139.25 ft^3 or about 140 ft^3

33. B

If we know the coordinates of two points on a line, we can find the slope (m) with the below formula:
$m = (y_2 - y_1)/(x_2 - x_1)$ where (x_1, y_1) represent the coordinates of one point and (x_2, y_2) the other.

In this question:

$(-4, -4) : x_1 = -4, y_1 = -4$

$(-1, 2) : x_2 = -1, y_2 = 2$

Inserting these values into the formula:

$m = (2 - (-4))/(-1 - (-4)) = (2 + 4)/(-1 + 4) = 6/3$ …
Simplifying by 3:

$m = 2$

34. C

Pythagorean Theorem:
$(\text{Hypotenuse})^2 = (\text{Perpendicular})^2 + (\text{Base})^2$

$h^2 = a^2 + b^2$

Given: $3^2 + 4^2 = h^2$

$h^2 = 9 + 16$

$h = \sqrt{25}$

$h = 5$

35. C

Comparing respective sides, ABCD, KLMN, WXYZ are similar.

36. A

Yes the triangles are congruent.

37. A

The triangles are congruent. This is a case of ASA:

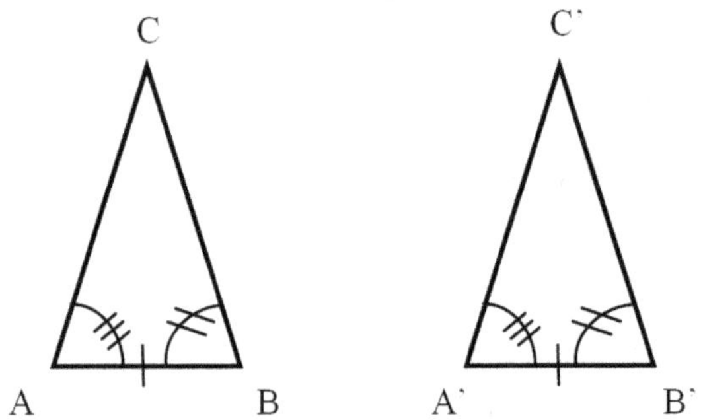

38. Yes the triangles are congruent

We mark the elements that are the same:

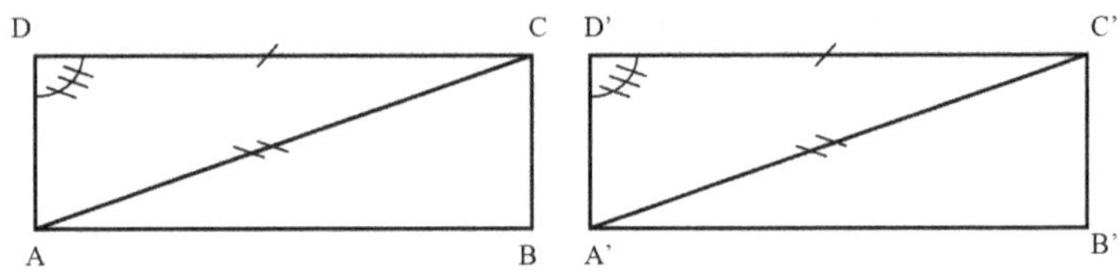

Angles at points D and D' are both right, because ABCD and A'B'C'D' are rectangles, so these 2 angles are equal. For triangles ACD and A'C'D' we have a case of SSA, so we have:

ACD ≈ A'C'D'

This means that sides AD and A'D' are equal. So, we can conclude that these 2 rectangles have all the same angles and same sides, so they are congruent.

39. Yes the triangles are congruent
CD = C'D'
AD = A'D'
Angle DCB = Angle D'C'B'

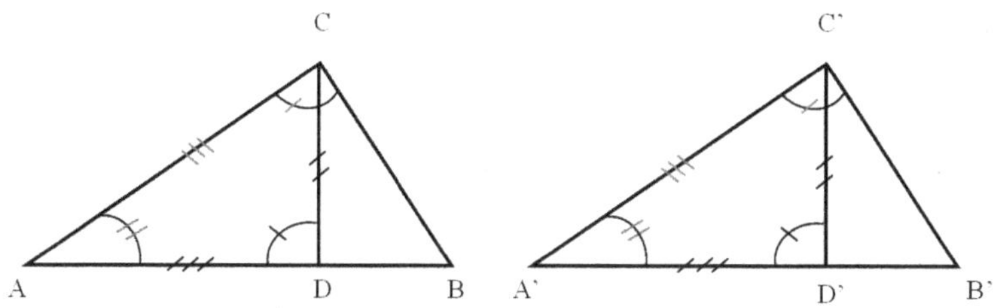

40. A
The triangles DEC and DAB are similar since AB is parallel to EC. Then, we have the similarity ratio:

|DE| / |DA| = 3/(3 + 5) = 3/8

The area of a triangle is found by base * height / 2. The similarity ratio 3/8 is valid for both the heights and bases of triangles DEC and DAB. This means that, the proportion of areas of DEC and DAB is the square of the similarity ratio:

A(DEC) / A(DAB) = $(3/8)^2$ = 9/64

A(ABCE) = A(DAB) - A(DEC)

Then;

A(DEC) / A(ABCE) = 9 / (64 - 9) = 9/55

Algebra

THE BASIC ALGEBRA SECTION COVERS THE FOLLOWING:

- Ratio and proportion
- Linear equations with 1 and 2 variables
- Quadratics
- Real-world quadratic questions
- Identify quadratic equations from graphs
- Identify linear equations from graphs
- Polynomials
- Solve Geometric problems with Algebra

Solving One-Variable Linear Equations

Linear equations with variable x is an equation with the following form:

$$ax = b$$

where a and b are real numbers. If a=0 and b is different from 0, then the equation has no solution.

Let's solve one simple example of a linear equation with one variable:

$$4x - 2 = 2x + 6$$

When given this type of equation, move variables to one side, and real numbers to the other. Always remember: if you are changing sides, you are changing signs. Move all variables to the left, and real numbers to the right:

$4x - 2 = 2x + 6$

$4x - 2x = 6 + 2$
$2x = 8$
$x = 8/2$
$x = 4$

When 2x goes to the left it becomes -2x, and -2 goes to the right and becomes +2. After calculations, we find that x is 4, which is a solution of our linear equation.

Let's solve a little more complex linear equation:

$2x - 6/4 + 4 = x$
$2x - 6 + 16 = 4x$
$2x - 4x = -16 + 6$
$-2x = -10$
$x = -10/-2$
$x = 5$

We multiply whole equation by 4, to lose the fractional line. Now we have a simple linear equation. If we change sides, we change the signs.

Solving Two-Variable Linear Equations

If we have 2 or more linear equations with 2 or more variables, then we have a system of linear equations. The idea here is to express one variable using the other in one equation, and then use it in the second equation, so we get a linear equation with one variable. Here is an example:
$x - y = 3$
$2x + y = 9$
From the first equation, we express y using x.

$y = x - 3$

In the second equation, we write x-3 instead of y. And there we get a linear equation with one variable x.

$2x + x - 3 = 9$
$3x = 9 + 3$
$3x = 12$
$x = 12/3$
$x = 4$

Now that we found x, we can use it to find y.

$y = x - 3$
$y = 4 - 3$
$y = 1$

So, the solution of this system is (x,y) = (4,1).

Let's solve one more system using a different method:

Solve:

$5x - 3y = 17$
$x + 3y = 11$

$5x - 3y + x + 3y = 17 - 11$

Notice that we have -3y in the first equation and +3y in the second. If we add these 2, we get zero, which means we lose variable y. So, we add these 2 equations and we get a linear equation with one variable.

$6x = 6$
$x = 1$
Now that we have x, we use it to find y.

$5 - 3y = 17$
$-3y = 17 - 5$

-3y = 12
y = 12/(-3)
y = -4

Basic Algebra Solutions - Video Tutorials

https://test-preparation.ca/algebra-practice-questions/

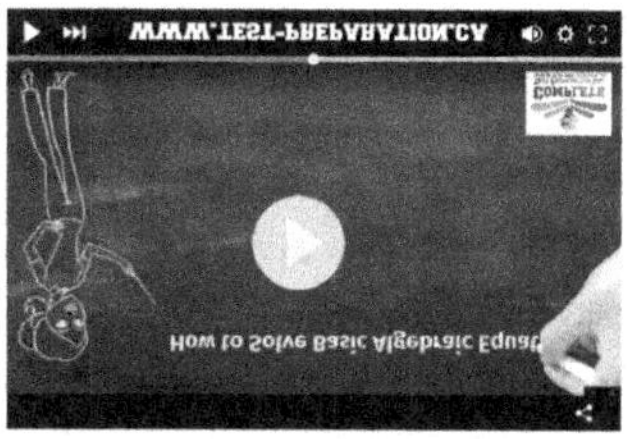

1. 5x + 3 = 7x – 1. Find x

a. 1/3
b. ½
c. 1
d. 2

https://youtu.be/3v6_GBUSuE4

Solution

1. D
5x + 3 = 7x – 1
first we need to collect like terms
second is to undo addition and subtraction
third and the last is to undo multiplication and division
so again the first step is to put 5x and 7x in one side = 7x-5x
second put 3 and 1 in one side = 7x – 5x = 3 + 1
now we will solve the answer = 2x = 4
now we will divide by 2 to the both sides
2x/2 = 4/2
the answer now is x=2

2. 5x + 2(x + 7) = 14x – 7. Find x

a. 1
b. 2
c. 3
d. 4

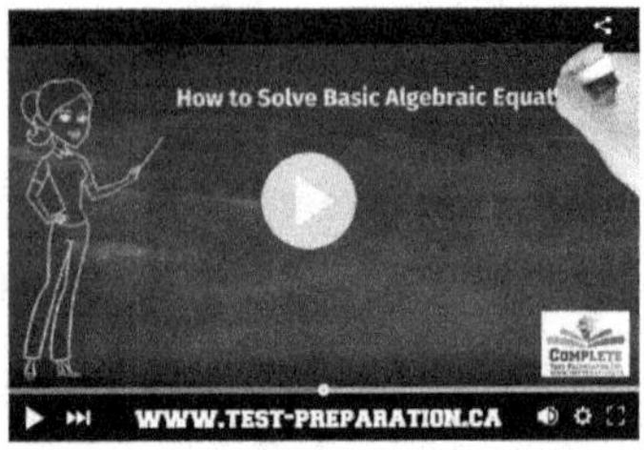

https://youtu.be/xRccgQvkz_8

Solution

2. C
5x + 2(x + 7) = 14x – 7
5x + 2x + 14 = 14x – 7
7x + 14 = 14x – 7
7x – 14x = -14 – 7
-7x = -21
x = 3

3. 12t – 10 = 14t + 2. Find t

a. -6
b. -4
c. 4
d. 6

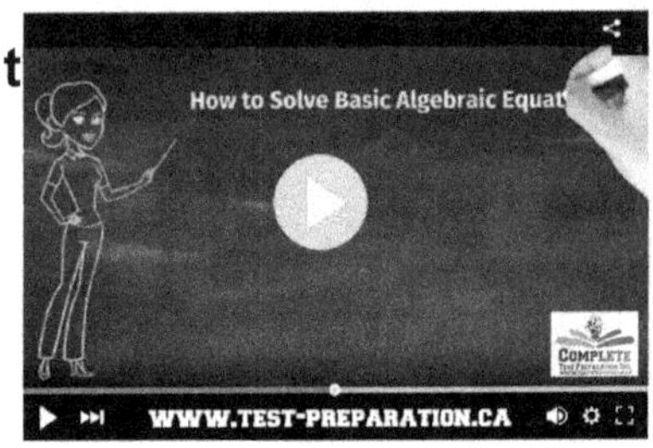

https://youtu.be/d9dsKx7U0Tg

3. A

$12t - 10 = 14t + 2$
$12t - 14t = 10 + 2$
$-2t = 12$
$-t = 6$
$t = -6$

4. 5(z + 1) = 3(z + 2) + 11 Solve for Z

a. 2
b. 4
c. 6
d. 12

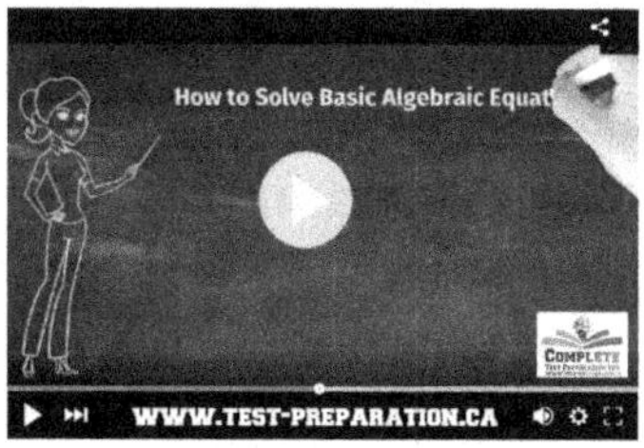

Solution

https://youtu.be/jWGAPCV1zRI

4. C

$5(z + 1) = 3(z + 2) + 11$. $Z=?$
$5z + 5 = 3z + 6 + 11$
$5z + 5 = 3z + 17$
$5z = 3Z + 17 - 5$
$5z - 3z = 12$
$2z = 12$
$z = 6$

Common Mistakes Answering Linear Equation Questions

Misinterpreting the question or setting up the equation Incorrectly

Failing to properly understand what the question is asking. Confusing linear equations with other types of algebra.

Misreading the problem and setting up the wrong equation.

Basic arithmetic errors: Addition, subtraction, multiplication, or division errors.
Basic Math Practice
https://test-preparation.ca/basic-math-practice-2/

Incorrectly applying the order of operations (PEMDAS). Order of Operation Practice
https://test-preparation.ca/order-of-operation/

Solving for the Wrong Variable:

Sign Errors: Incorrectly using positive and negative signs. Failing to change the sign when performing basic operations.

Fractions and Decimals: Incorrectly converting between fractions, decimals, and percentages.
Factions, decimals and percent practice

.Incorrect Distribution: Mistakes when applying the distributive property. Forgetting to distribute a negative sign through parentheses.

Incorrectly Combining Like Terms: Failing to properly combine like terms.

Mistaking terms that are not alike for like terms.

Skipping Steps: Skipping steps in the process, leading to errors in the solution.

Not showing work, making it hard to identify where a mistake was made.

Checking Solutions: Forgetting to plug the solution back into the original equation to check its correctness.

Assuming an answer is correct without verification.

Mistakes Graphing: Errors plotting or connecting points. Misinterpreting the slope and intercept in the line equation.

Time Management: Spending too much time on one problem and rushing through others. Not allocating enough time to review and check answers.
Time Management on a Test
https://test-preparation.ca/test-tactics-the-time-wise-approach/

Anxiety and Pressure: Feeling overwhelmed or stressed causing incorrect answers. Making careless mistakes due to test anxiety. Test Anxiety Tips
https://test-preparation.ca/how-to-overcome-test-anxiety/

Simplifying Polynomials

Let's say we are given some expression with one or more variables, where we have to add, subtract and multiply polynomials. We do the calculations with variables and constants and then we group the variables with the appropriate degrees. As a result, we would get a polynomial. This process is called simplifying polynomials, where we go from a complex expression to a simple polynomial.

Example:

Simplify the following expression and arrange the degrees from bigger to smaller:

$4 + 3x - 2x^2 + 5x + 6x^3 - 2x^2 + 1 = 6x^3 - 4x^2 + 8x + 5$

We can have more complex expressions such as:

$(x + 5)(1 - x) - (2x - 2) = x - x^2 + 5 - 5x - 2x + 2 = -x^2 - 6x + 7$

Here, first we multiply the polynomials and then we subtract the result and the third polynomial.

Factoring Polynomials

If we have a polynomial that we want to write as multiplication of a real number and a polynomial or as a multiplication of 2 or more polynomials, then we are dealing with factoring polynomials.

Let's see an example for a simple factoring:

$12x^2 + 6x - 4 =$
$2 * 6x^2 + 2 * 3x - 2 * 2 =$
$2(6x^2 + 3x - 2)$

We look at every polynomial member as a product of a real number and a variable. Notice that all real numbers in the polynomial are even, so they have the same number (factor). We pull out that 2 in front of the polynomial, and we write what is left.

What if have a more complex case, where we can't find a factor that is a real number? Here is an example:

$x^2 - 2x + 1 =$
$x^2 - x - x + 1 =$
$x(x - 1) - (x - 1) =$
$(x - 1) (x - 1)$

We can write -2x as –x-x . Now we group first 2 members and we see that they have the same factor x, which we can pull in front of them. For the other 2 members, we pull the minus in front of them, so we can get the same binomial that we got with the first 2 members. Now we have that this binomial is the factor for x(x-1) and (x-1).

If we pull x-1 in front (underlined), from the first member we are left with x, and from the second we have -1.
And that is how we transform a polynomial into a product of 2 polynomials (in this case binomials).

Quadratic Equations

A. Factoring

Quadratic equations are usually called second degree equations, which mean that the second degree is the highest degree of the variable that can be found in the quadratic equation. The form of these equations is:

$ax^2 + bx + c = 0$

where a, b and c are some real numbers.

One way for solving quadratic equations is the factoring method, where we transform the quadratic equation into a product of 2 or more polynomials. Let's see how that works in one simple example:

$x2 + 2x = 0$
$x(x + 2) = 0$
$(x = 0) V (x + 2 = 0)$
$(x = 0 V (x + -2)$

Notice that here we don't have parameter c, but this is still a quadratic equation, because we have the second degree of variable x. Our factor here is x, which we put in front, and we are left with x+2. The equation is equal to 0, so either x or x+2 are 0, or both are 0.
So, our 2 solutions are 0 and -2.

B. Quadratic formula

If we are unsure how to rewrite quadratic equations so we can solve it using factoring method, we can use the formula for quadratic equation:

$$x_{1,2} = \frac{-b \pm \sqrt{b^2 - 4ac}}{2a}$$

We write $x_{1,2}$ because it represents 2 solutions of the equation. Here is one example:

$$3x^2 - 10x + 3 = 0$$

$$x_{1,2} = \frac{-b \pm \sqrt{b^2 - 4ac}}{2a}$$

$$x_{1,2} = \frac{-(-10) \pm \sqrt{(-10)^2 - 4 \cdot 3 \cdot 3}}{2 \cdot 3}$$

$$x_{1,2} = \frac{10 \pm \sqrt{100 - 36}}{6}$$

$$x_{1,2} = \frac{10 \pm \sqrt{64}}{6}$$

$$x_{1,2} = \frac{10 \pm 8}{6}$$

$$x_1 = \frac{10 + 8}{6} = \frac{18}{6} = 3$$

$$x_2 = \frac{10 - 8}{6} = \frac{2}{6} = \frac{1}{3}$$

We see that a is 3, b is -10 and c is 3.
We use these numbers in the equation and do some calculations.

Notice that we have + and -, so x_1 is for + and x_2 is for -, and that's how we get 2 solutions.

Quadratic Word Problems

Some real life problems can be solved using quadratic equations. Always try to read carefully so you can write the math problem correctly. Even some geometrical problems can be solved using quadratic equations. Let's solve one real life problem with quadratics:

The distance between 2 cities is 588 kilometers. Train A travels that distance 2 hours and 20 minutes less than train B. How fast are the trains traveling if they differ by 21 km/h?

We first write 2 hours and 20 minutes as a fraction: 20 minutes is one third of the one full hour, so we have:

2 1/3 h = 7/3 h

If we denote with S the distance of 588 km, and V1 is the speed of the train A and V2 is the speed of the train B, and times of the travel are t1 and t2, respectively. Now we can write:

$S = 588$
$V_1 - V_2 = 21 \rightarrow V_2 = V_1 - 21$

$t_1 + 7/3 = t_2$
$V_1 = S/\ t_1 \rightarrow V_1\ t_1 \rightarrow 588 = V_1\ t_1 \rightarrow t_1\ 588\ /\ V_1$

$V_2 = S/\ t_2 \rightarrow S = V_2\ t_2 \rightarrow 588 = (V_1 - 21)(t_1 + 7/3)$

$588 = (V_1 - 21)\ (588/\ V_1 + 7/3)$

$588 = 588 + 7/3\ V_1 - 21 * 588/\ V_1 - 21\ 7/3$

$0 = 7/x\ V_1 - 3 * 588/v1 - 7/\text{-}3\ V_1$
$0 = V_1^2 - 5292 - 21\ V_1$
$v_1^2 - 21V_1 - 5292 = 0$

$V_{1,2} = (21 \pm \sqrt{441 + 4 * 5292})\ /\ 2$

$V_{1,2} = (21 + 147)/2$

$V_1 = 84$ km/h

$V_2 = 84 - 21 = 63$ km/h

Quadratic Geometry Problems

If length of a hypotenuse of a right triangle is 5, and sum of its legs is 7, find the lengths of its legs.

$a^2 + b^2 = c^2 = 5^2 = 25$

$a + b = 7 \rightarrow a = 7 - b$

$(7 - b)^2 + b^2 = 25$

$49 - 14b + b^2 + b^2 = 25$

$2b^2 - 14b + 24 = 0$

$b^2 - 7b + 24 = 0$

$b_{1,2} = (7 \pm \sqrt{(49 - (4 * 12)}) / 2$

$b_{1,2} = 7 \pm \sqrt{(49 - 48)}/2$

$b_{1,2} = (7 \pm 1)/2$

$b_1 = 4$

$b_2 = 3$

$a_1 = 7 - 4 = 3$
$a_2 = 7 - 3 = 4$

Answer Sheet

	A	B	C	D	E		A	B	C	D	E
1	○	○	○	○	○	26	○	○	○	○	○
2	○	○	○	○	○	27	○	○	○	○	○
3	○	○	○	○	○	28	○	○	○	○	○
4	○	○	○	○	○	29	○	○	○	○	○
5	○	○	○	○	○	30	○	○	○	○	○
6	○	○	○	○	○	31	○	○	○	○	○
7	○	○	○	○	○	32	○	○	○	○	○
8	○	○	○	○	○	33	○	○	○	○	○
9	○	○	○	○	○	34	○	○	○	○	○
10	○	○	○	○	○	35	○	○	○	○	○
11	○	○	○	○	○	36	○	○	○	○	○
12	○	○	○	○	○	37	○	○	○	○	○
13	○	○	○	○	○	38	○	○	○	○	○
14	○	○	○	○	○	39	○	○	○	○	○
15	○	○	○	○	○	40	○	○	○	○	○
16	○	○	○	○	○	41	○	○	○	○	○
17	○	○	○	○	○	42	○	○	○	○	○
18	○	○	○	○	○	43	○	○	○	○	○
19	○	○	○	○	○	44	○	○	○	○	○
20	○	○	○	○	○	45	○	○	○	○	○
21	○	○	○	○	○	46	○	○	○	○	○
22	○	○	○	○	○	47	○	○	○	○	○
23	○	○	○	○	○	48	○	○	○	○	○
24	○	○	○	○	○	49	○	○	○	○	○
25	○	○	○	○	○	50	○	○	○	○	○

Basic Algebra Practice

1. Solve the linear equation: $-x - 7 = -3x - 9$

a. -1

b. 0

c. 1

d. 2

2. Solve the system: $4x - y = 5 \quad x + 2y = 8$

a. (3,2)

b. (3,3)

c. (2,3)

d. (2,2)

3. Simplify the following expression:

$3x^3 + 2x^2 + 5x - 7 + 4x^2 - 5x + 2 - 3x^3$

a. $6x^2 - 9$

b. $6x^2 - 5$

c. $6x^2 - 10x - 5$

d. $6x^2 + 10x - 9$

4. Find 2 numbers that sum to 21 and the sum of the squares is 261.

a. 14 and 7

b. 15 and 6

c. 16 and 5

d. 17 and 4

5. Using the factoring method, solve the quadratic equation: $x^2 + 4x + 4 = 0$

a. 0 and 1

b. 1 and 2

c. 2

d. -2

6. Using the quadratic formula, solve the quadratic equation: $x - 31/x = 0$

a. $-\sqrt{13}$ and $\sqrt{13}$

b. $-\sqrt{31}$ and $\sqrt{31}$

c. $-\sqrt{31}$ and $2\sqrt{31}$

d. $-\sqrt{3}$ and $\sqrt{3}$

7. Using the factoring method, solve the quadratic equation: $2x^2 - 3x = 0$

a. 0 and 1.5

b. 1.5 and 2

c. 2 and 2.5

d. 0 and 2

8. Using the quadratic formula, solve the quadratic equation: $x^2 - 9x + 14 = 0$

a. 2 and 7

b. -2 and 7

c. -7 and -2

d. -7 and 2

9. Solve the following equation $4(y + 6) = 3y + 30$

a. $y = 20$
b. $y = 6$
c. $y = 30/7$
d. $y = 30$

10. Using the factoring method, solve the quadratic equation: $x^2 - 5x - 6 = 0$

a. -6 and 1
b. -1 and 6
c. 1 and 6
d. -6 and -1

11. Factor the polynomial $x^3y^3 - x^2y^8$.

a. $x^2y^3(x - y^5)$
b. $x^3y^3(1 - y^5)$
c. $x^2y^2(x - y^6)$
d. $xy^3(x - y^5)$

12. Find the solution for the following linear equation: $5x/2 = (3x + 24)/6$

a. -1
b. 0
c. 1
d. 2

13. Solve the system, if a is some real number:

$ax + y = 1$
$x + ay = 1$

a. (1,a)

b. (1/a + 1, 1)

c. (1/(a + 1), 1/(a + 1))

d. (a, 1/a + 1)

14. Solve $3(x + 2) - 2(1 - x) = 4x + 5$

a. -1

b. 0

c. 1

d. 2

15. Simplify $3x^a + 6a^x - x^a + (-5a^x) - 2x^a$

a. $a^x + x^a$

b. $a^x - x^a$

c. a^x

d. x^a

16. A map uses a scale of 1:100,000. How much distance on the ground is 3 inches on the map if the scale is in inches?

a. 13 inches

b. 300,000 inches

c. 30,000 inches

d. 333.999 inches

17. Using the quadratic formula, solve the quadratic equation: $0.9x^2 + 1.8x - 2.7 = 0$

a. 1 and 3

b. -3 and 1

c. -3 and -1

d. -1 and 3

18. Find x and y from the following system of equations:

$(4x + 5y)/3 = ((x - 3y)/2) + 4$
$(3x + y)/2 = ((2x + 7y)/3) -1$

a. (1, 3)

b. (2, 1)

c. (1, 1)

d. (0, 1)

19. Using the factoring method, solve the quadratic equation: $x^2 + 12x - 13 = 0$

c. a. -13 and 1

d. b. -13 and -1

e. c. 1 and 13

f. d. -1 and 13

20. Using the quadratic formula, solve the quadratic equation: $((x^2 + 4x + 4) + (x^2 - 4x + 4)) / (x^2 - 4) = 0$.

a. It has infinite numbers of solutions

b. 0 and 1

c. It has no solutions

d. 0

21. Turn the following expression into a simple polynomial: $5(3x^2 - 2) - x^2(2 - 3x)$

a. $3x^3 + 17x^2 - 10$

b. $3x^3 + 13x^2 + 10$

c. $-3x^3 - 13x^2 - 10$

d. $3x^3 + 13x^2 - 10$

22. Solve $(x^3 + 2)(x^2 - x) - x^5$.

a. $2x^5 - x^4 + 2x^2 - 2x$

b. $-x^4 + 2x^2 - 2x$

c. $-x^4 - 2x^2 - 2x$

d. $-x^4 + 2x^2 + 2x$

23. $9ab^2 + 8ab^2 =$

a. ab^2

b. $17ab^2$

c. 17

d. $17a^2b^2$

24. Factor the polynomial $x^2 - 7x - 30$.

a. $(x + 15)(x - 2)$

b. $(x + 10)(x - 3)$

c. $(x - 10)(x + 3)$

d. $(x - 15)(x + 2)$

25. If a and b are real numbers, solve the following equation: $(a + 2)x - b = -2 + (a + b)x$

a. -1

b. 0

c. 1

d. 2

26. Turn the following expression into a simple polynomial: $1 - x(1 - x(1 - x))$

a. $x^3 + x^2 - x + 1$

b. $-x^3 - x^2 + x + 1$

c. $-x^3 + x^2 - x + 1$

d. $x^3 + x^2 - x - 1$

27. $7(2y + 8) + 1 – 4(y + 5) =$

a. $10y + 36$

b. $10y + 77$

c. $18y + 37$

d. $10y + 37$

28. Richard gives 's' amount of salary to each of his 'n' employees weekly. If he has 'x' amount of money then how many days he can employ these 'n' employees.

a. sx/7n

b. 7x/nx

c. nx/7s

d. 7x/ns

29. Factor the polynomial $x^2 - 3x - 4$.

a. $(x + 1)(x - 4)$
b. $(x - 1)(x + 4)$
c. $(x - 1)(x - 4)$
d. $(x + 1)(x + 4)$

30. Using the quadratic formula, solve the quadratic equation:

$(a^2 - b^2)x^2 + 2ax + 1 = 0$

a. $a/(a + b)$ and $b/(a + b)$
b. $1/(a + b)$ and $a/(a + b)$
c. $a/(a + b)$ and $a/(a - b)$
d. $-1/(a + b)$ and $-1/(a - b)$

31. Turn the following expression into a simple polynomial: $(a + b)(x + y) + (a - b)(x - y) - (ax + by)$

a. $ax + by$
b. $ax - by$
c. $ax^2 + by^2$
d. $ax^2 - by^2$

32. The area of a rectangle is 20 cm². If one side increases by 1 cm and other by 2 cm, the area of the new rectangle is 35 cm². Find the sides of the original rectangle.

a. (4,8)
b. (4,5)
c. (2.5,8)
d. b and c

33. Find the x-intercepts of the quadratic function $f(x) = (x - 5)^2 - 9$.

a. {2,4}
b. {2,8}
c. {4,8}
d. {1,2}

34. In a store, the price of t-shirts and pants are constant. If John buys 4 t-shirts and 5 pair of pants, he pays $51. If he buys 7 t-shirts and 3 pair of pants, then he pays $49. Find the difference between the price of one pair of pants and one t-shirt.

a. 0
b. 3
c. 7
d. 12

35. Which of the following graphs represent the equation $4x - y = 6$?

a.

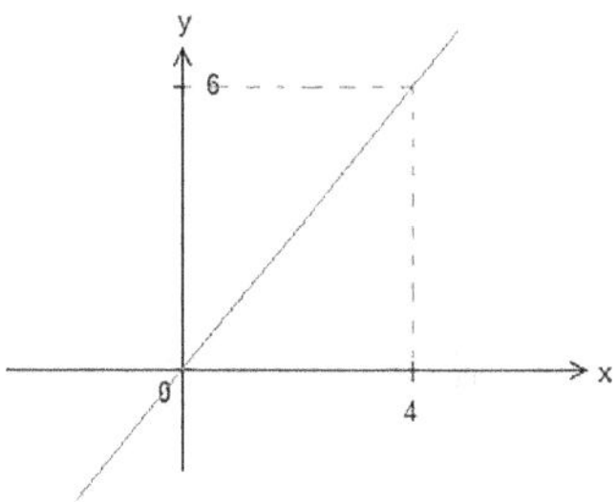

b.

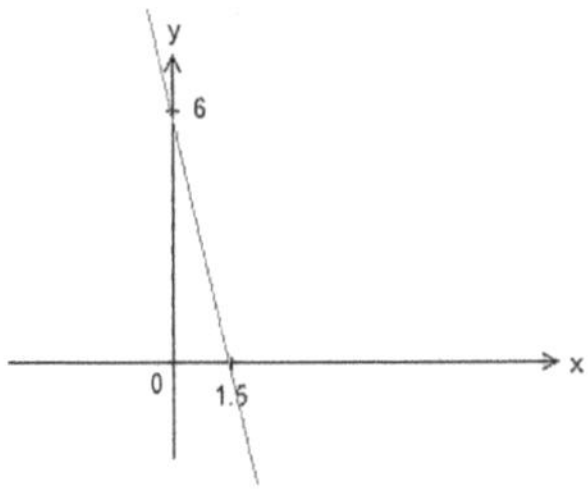

c.

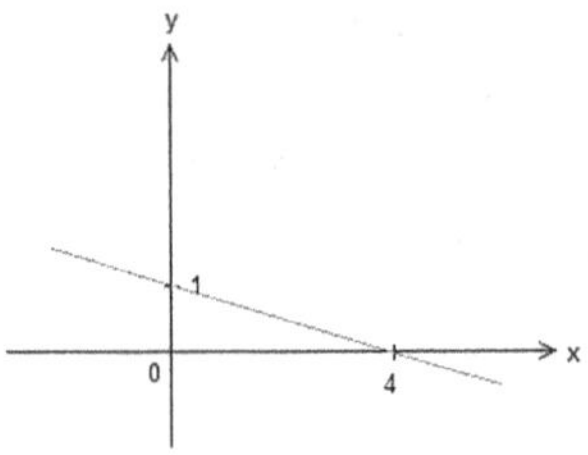

d.

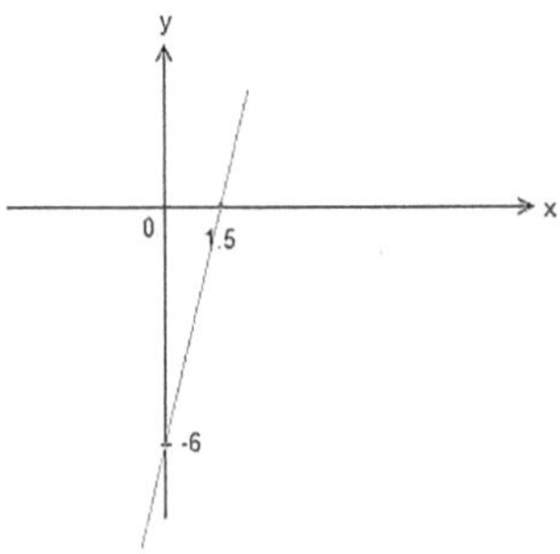

36. A number is increased by 2 and then multiplied by 3. The result is 24. What is this number?

a. 4
b. 6
c. 8
d. 10

37. My father's age divided by 5 is equal to my brother's age divided by 3. My brother is 3 years older than me. My father's age is 3 less than 2 times my age. How old is my father?

a. 34

b. 45

c. 56

d. 61

38. (x – 2) / 4 – (3x + 5) / 7 = - 3, x=?

a. 6

b. 7

c.10

d.13

39. The inner angles of a triangle are given as x + 20, 3x – 10 and 8x + 50. Find the difference between the smallest and the largest angles.

a. 65

b. 75

c. 110

d. 150

40. The width of a rectangle is two thirds of the length. The perimeter of this rectangle is 150 cm. Find the length of this shape.

a. 30 cm

b. 45 cm

c. 60 cm

d. 75 cm

41. The volume of a sphere with radius r is equal to the volume of a cylinder with radius 3r and height h. What is r/h equal to?

a. 9/4

b. 9/2

c. 5

d. 27/4

42. The area of a triangle is equal to 32 cm^2 and the height of this triangle is 4 cm less than 3 times the base. What is the length of the height?

a. 12

b. 16

c. 18

d. 24

43. What is the sum of possible integers for x satisfying $|5x - 3| \leq 7$?

a. 1

b. 2

c. 3

d. 5

44. What is the multiplication of possible x values satisfying $|7 – 2x| = 9$?

a. -8

a. -1

a. 4

a. 16

45. Find the solution set for $|x - 5| + 8 < 4$.

a. { }

b. {-3, 4}

c. {1, 3}

d. {1, 4}

Answer Key

1. A

We should collect similar terms on the same side. Here, we can collect x terms on left side, and the constants on the right side:

$- x - 7 = - 3x - 9$ …. Let us add 3x to both sides:

$- x - 7 + 3x = - 3x - 9 + 3x$

$2x - 7 = - 9$ … Now, we can add + 7 to both sides:

$2x - 7 + 7 = - 9 + 7$

$2x = - 2$ … Dividing both sides by 2 gives us the value of x:

$x = -2/2$

$x = -1$

2. C

First, we need to write two equations separately:
$4x - y = 5$ (I)

$x + 2y = 8$ (II) … Here, we can use two ways to solve the system. One is substitution method, the other one is linear elimination method:

1. Substitution Method

Equation (I) gives us that $y = 4x - 5$. We insert this value of y into equation (II):

$x + 2(4x - 5) = 8$

$x + 8x - 10 = 8$

$9x - 10 = 8$

$9x = 18$

$x = 2$

Bu knowing $x = 2$, we can find the value of y by inserting $x = 2$ into either of the equations. Let us choose equation (I):

$4(2) - y = 5$

$8 - y = 5$

$8 - 5 = y$

$y = 3 \rightarrow$ solution is (2, 3)

2. Linear Elimination Method:

2•/ $4x - y = 5$ … by multiplying equation (I) by 2, we see that -2y will form; and y terms

$x + 2y = 8$ … will be eliminated when summed with +2y in equation (II):

2•/ $4x - y = 5$

$+ \quad \underline{x + 2y = 8}$

$8x - 2y = 10$

$+ \quad \underline{x + 2y = 8}$ … Summing side by side:

$8x + x - 2y + 2y = 10 + 8$ … -2y and +2y eliminate each other:

$9x = 18$

$x = 2$

By knowing $x = 2$, we can find the value of y by inserting $x = 2$ into either of the equations. Let us choose equation (I):

$4(2) - y = 5$

$8 - y = 5$

$8 - 5 = y$

$y = 3 \rightarrow$ solution is (2, 3)

3. B

$3x^3 + 2x^2 + 5x - 7 + 4x^2 - 5x + 2 - 3x^3$ … write similar terms together:

$= 3x^3 - 3x^3 + 2x^2 + 4x^2 + 5x - 5x - 7 + 2$ … operate within the same terms. $3x^3$ and $- 3x^3$, 5x and -5x cancel:

$= 6x^2 - 5$

4. B

There are two statements made. This means that we can write two equations according to these statements:

The sum of two numbers are 21: $x + y = 21$

The sum of the squares is 261: $x^2 + y^2 = 261$

We are asked to find x and y.

Since we have the sums of the numbers and the sums of their squares; we can use the square formula of x + y, that is:

$(x + y)^2 = x^2 + 2xy + y^2$ … Here, we can insert the known values $x + y$ and $x^2 + y^2$:

$(21)^2 = 261 + 2xy$ … Arranging to find xy:

$441 = 261 + 2xy$

$441 - 261 = 2xy$

$180 = 2xy$

$xy = 180/2$

$xy = 90$

We need to find two numbers which multiply to 90. Checking the answer choices, we see that in (b), 15 and 6 are given. 15•6 = 90. Also their squares sum up to 261 ($15^2 + 6^2 = 225 + 36 = 261$). So these two numbers satisfy the equation.

5. D
$x^2 + 4x + 4 = 0$ … We try to separate the middle term 4x to find common factors with x2 and 4 separately:

$x^2 + 2x + 2x + 4 = 0$ … Here, we see that x is a common factor for x^2 and 2x, and 2 is a common factor for 2x and 4:

$x(x + 2) + 2(x + 2) = 0$ … Here, we have x times x + 2 and 2 times x + 2 summed up. This means that we have x + 2 times x + 2:

$(x + 2)(x + 2) = 0$

$(x + 2)^2 = 0$ … This is true if only if x + 2 is equal to zero.

$x + 2 = 0$

$x = -2$

6. B
To solve the equation, first we need to arrange it to appear in the form $ax^2 + bx + c = 0$ by removing the denominator:

$x - 31/x = 0$ … First, we enlarge the equation by x:

$x \cdot x - 31 \cdot x/x = 0$

$x^2 - 31 = 0$

The quadratic formula to find the roots of a quadratic equation is:

$x_{1,2} = (-b \pm \sqrt{\Delta}) / 2a$ where $\Delta = b^2 - 4ac$ and is called the discriminant of the quadratic equation.

In our question, the equation is $x^2 - 31 = 0$. By remembering the form $ax^2 + bx + c = 0$:

$a = 1, b = 0, c = -31$

So, we can find the discriminant first, and then the roots of the equation:

$\Delta = b^2 - 4ac = 0^2 - 4 \cdot 1 \cdot (-31) = 124$

$x_{1,2} = (-b \pm \sqrt{\Delta}) / 2a = (\pm\sqrt{124}) / 2 = (\pm\sqrt{4 \cdot 31}) / 2 = (\pm 2\sqrt{31}) / 2$ ... Simplifying by 2:

$x_{1,2} = \pm\sqrt{31}$ ... This means that the roots are $\sqrt{31}$ and $-\sqrt{31}$.

7. A
$2x^2 - 3x = 0$ ... we see that both of the terms contain x; so we can take it out as a factor:

$x(2x - 3) = 0$ ... two terms are multiplied and the result is zero. This means that either of the terms or, both can be equal to zero:

$x = 0$ ... this is one solution

$2x - 3 = 0 \rightarrow 2x = 3 \rightarrow x = 3/2 \rightarrow x = 1.5$ ... this is the second solution.

So, the solutions are 0 and 1.5.

8. A
To solve the equation, we need the equation in the form $ax^2 + bx + c = 0$.

$x^2 - 9x + 14 = 0$ is already in this form.

The quadratic formula to find the roots of a quadratic equation is:

$x_{1,2} = (-b \pm \sqrt{\Delta}) / 2a$ where $\Delta = b^2 - 4ac$ and is called the discriminant of the quadratic equation.

In our question, the equation is $x^2 - 9x + 14 = 0$. By remembering the form $ax^2 + bx + c = 0$:

$a = 1, b = -9, c = 14$

So, we can find the discriminant first, and then the roots of the equation:

$\Delta = b^2 - 4ac = (-9)^2 - 4 \times 1 \times 14 = 81 - 56 = 25$

$x_{1,2} = (-b \pm \sqrt{\Delta}) / 2a = (-(-9) \pm \sqrt{25}) / 2 = (9 \pm 5) / 2$

This means that the roots are,

$x_1 = (9 - 5) / 2 = 2$ and $x_2 = (9 + 5) / 2 = 7$

9. B
$4y + 24 = 3y + 30, = 4y – 3y + 24 = 30, = y + 24 = 30, = y = 30 – 24, = y = 6$

10. B
$x^2 - 5x - 6 = 0$

We try to separate the middle term -5x to find common factors with x^2 and -6 separately:

$x^2 - 6x + x - 6 = 0$ … Here, we see that x is a common factor for x^2 and -6x:

$x(x - 6) + x - 6 = 0$ … Here, we have x times x - 6 and 1 time x - 6 summed up. This means that we have x + 1 times x - 6:

$(x + 1)(x - 6) = 0$ ... This is true when either or both of the expressions in the parenthesis are equal to zero:

$x + 1 = 0$ ... $x = -1$

$x - 6 = 0$ ... $x = 6$

-1 and 6 are the solutions for this quadratic equation.

11. A

We need to find the greatest common divisor of the two terms to factor the expression. We should remember that if the bases of exponent numbers are the same, the multiplication of two terms is found by summing the powers and writing on the same base. Similarly; when dividing, the power of the divisor is subtracted from the power of the divided.

Both x^3y^3 and x^2y^8 contain x^2 and y^3. So;

$x^3y^3 - x^2y^8 = x \cdot x^2y^3 - y^5 \cdot x^2y^3$ ... We can carry x^2y^3 out as the factor:

$= x^2y^3(x - y^5)$

12. D

Our aim to collect the knowns on one side and the unknowns (x terms) on the other side:

$5x/2 = (3x + 24)/6$ ... First, we can simplify the denominators of both sides by 2:

$5x = (3x + 24)/3$ ... Now, we can cross multiply:

$15x = 3x + 24$

$15x - 3x = 24$

$12x = 24$

$x = 24/12 = 2$

13. C
Solving the system means finding x and y. Since we also have a in the system, we will find x and y depending on a.

We can obtain y by using the equation $ax + y = 1$:

$y = 1 - ax$ ... Then, we can insert this value into the second equation:

$x + a(1 - ax) = 1$

$x + a - a^2x = 1$

$x - a^2x = 1 - a$

$x(1 - a^2) = 1 - a$ ... We need to obtain x alone:

$x = (1 - a)/(1 - a^2)$ ... Here, $1 - a^2 = (1 - a)(1 + a)$ is used:

$x = (1 - a)/((1 - a)(1 + a))$ ... Simplifying by $(1 - a)$:

$x = 1/(a + 1)$ ... Now we know the value of x. By using either of the equations, we can find the value of y. Let us use $y = 1 - ax$:

$y = 1 - a \cdot 1/(a + 1)$

$y = 1 - a/(a + 1)$ ... By writing on the same denominator:

$y = ((a + 1) - a)/(a + 1)$

$y = (a + 1 - a)/(a + 1)$ ... a and -a cancel each other:

$y = 1/(a + 1)$ ... x and y are found to be equal.

The solution of the system is $(1/(a + 1), 1/(a + 1))$

14. C
To solve the linear equation, we operate the knowns and unknowns within each other and try to obtain x term (which is the unknown) alone on one side of the equation:

$3(x + 2) - 2(1 - x) = 4x + 5$ ... We remove the parenthesis by distributing the factors:

$3x + 6 - 2 + 2x = 4x + 5$

$5x + 4 = 4x + 5$

$5x - 4x = 5 - 4$

$x = 1$

15. C

Here, we use the commutative property of multiplication, meaning that $xa = ax$:

$3xa + 6ax - xa + (-5ax) - 2xa = 3ax + 6ax - ax - 5ax - 2ax$

$= (3 + 6 - 1 - 5 - 2)ax$

$= (9 - 8)ax$

$= ax$

16. B

1 inch on map = 100,000 inches on ground. So 3 inches on map = 3 x 100,000 = 300,000 inches on ground.

17. B

To solve the equation, we need the equation in the form ax2 + bx + c = 0.

$0.9x^2 + 1.8x - 2.7 = 0$ is already in this form.

The quadratic formula to find the roots of a quadratic equation is:

$x_{1,2} = (-b \pm \sqrt{\Delta}) / 2a$ where $\Delta = b^2 - 4ac$ and is called the discriminant of the quadratic equation.

In our question, the equation is $0.9x^2 + 1.8x - 2.7 = 0$. To eliminate the decimals, let us multiply the equation by 10: $9x^2 + 18x - 27 = 0$ ... This equation can be simplified by 9 since each term contains 9:

$x^2 + 2x - 3 = 0$

By remembering the form $ax^2 + bx + c = 0$:

$a = 1, b = 2, c = -3$

So, we can find the discriminant first, and then the roots of the equation:

$\Delta = b^2 - 4ac = (2)^2 - 4 \cdot 1 \cdot (-3) = 4 + 12 = 16$

$x_{1,2} = (-b \pm \sqrt{\Delta}) / 2a = (-2 \pm \sqrt{16}) / 2 = (-2 \pm 4) / 2$

This means that the roots are,

$x_1 = (-2 - 4)/2 = -3$ and $x_2 = (-2 + 4)/2 = 1$

18. C
First, we need to arrange the two equations to obtain the form $ax + by = c$. We see that there are 3 and 2 in the denominators of both equations. If we equate all at 6, then we can cancel all 6 in the denominators and have straight equations:

Equate all denominators at 6:

$2(4x + 5y)/6 = 3(x - 3y)/6 + 4 \cdot 6/6$ … Now we can cancel 6 in the denominators:

$8x + 10y = 3x - 9y + 24$ … We can collect x and y terms on left side of the equation:

$8x + 10y - 3x + 9y = 24$

$5x + 19y = 24$ … Equation (I)
Arrange the second equation:

3(3x + y)/6 = 2(2x + 7y)/6 - 1 * 6/6 … Now we can cancel 6 in the denominators:

9x + 3y = 4x + 14y - 6 … We can collect x and y terms on left side of the equation:

9x + 3y - 4x - 14y = -6

5x - 11y = -6 … Equation (II)

Now, we have two equations and two unknowns x and y. By writing the two equations one under the other and operating, we can find one unknown first, and find the other next:

5x + 19y = 24
-1/ 5x - 11y = -6 … If we substitute this equation from the upper one, 5x cancels -5x:

5x + 19y = 24

-5x + 11y = 6 … Summing side-by-side:

5x - 5x + 19y + 11y = 24 + 6

30y = 30 … Dividing both sides by 30:
y = 1

Inserting y = 1 into either of the equations, we can find the value of x. Choosing equation I:

5x + 19•1 = 24

5x = 24 - 19

5x = 5 … Dividing both sides by 5:
x = 1

So, $x = 1$ and $y = 1$ is the solution; shown as (1, 1).

19. A

$x^2 + 12x - 13 = 0$ ... We try to separate the middle term 12x to find common factors with x^2 and -13 separately:

$x^2 + 13x - x - 13 = 0$ ... Here, we see that x is a common factor for x^2 and 13x, and -1 is a common factor for -x and -13:

$x(x + 13) - 1(x + 13) = 0$ ... Here, we have x times x + 13 and -1 times x + 13 summed up. This means that we have x - 1 times x + 13:

$(x - 1)(x + 13) = 0$

This is true when either, or both, the expressions in the parenthesis are equal to zero:

$x - 1 = 0$ ... $x = 1$

$x + 13 = 0$ ... $x = -13$

1 and -13 are the solutions for this quadratic equation.

20. C

First, we need to simplify the equation:
$((x^2 + 4x + 4) + (x^2 - 4x + 4)) / (x^2 - 4) = 0$

$(x^2 + 4x + 4 + x^2 - 4x + 4) / (x^2 - 4) = 0$ ... 4x and -4x in the numerator cancel.

Note that $x^2 - 4$ is two square difference and is equal to $x^2 - 2^2 = (x - 2)(x + 2)$:

$(2x^2 + 8)/((x - 2)(x + 2)) = 0$

The denominator tells us that if x - 2 or x + 2 equals to zero, there will be no solution. So, we will need to eliminate $x = 2$ and $x = -2$ from our solution which will be found considering the numerator:

$2x^2 + 8 = 0$

$2(x^2 + 4) = 0$

$x^2 + 4 = 0$

$x^2 = -4$ ... We know that, a square cannot be equal to a negative number. Solution for the square root of -4 is not a real number, so this equation has no solution.

21. D
We need to distribute the factors to the terms inside the related parenthesis:

$5(3x^2 - 2) - x^2(2 - 3x) = 15x^2 - 10 - (2x^2 - 3x^3)$

$= 15x^2 - 10 - 2x^2 + 3x^3$

$= 3x^3 + 15x^2 - 2x^2 - 10$ ... similar terms written together to ease summing/substituting.

$= 3x^3 + 13x^2 - 10$

22. B
We need to distribute the factors to the terms inside the related parenthesis:

$(x^3 + 2)(x^2 - x) - x^5 = x^5 - x^4 + (2x^2 - 2x) - x^5$

$= x^5 - x^4 + 2x^2 - 2x - x^5$

$= x^5 - x^5 - x^4 + 2x^2 - 2x$ ... similar terms written together to ease summing/substituting.

$= -x^4 + 2x^2 - 2x$

23. B
To simplify the expression, we need to find common factors. We see that both terms contain the term ab^2. So, we can take this term out of each term as a factor:

$9ab^2 + 8ab^2 = (9 + 8)\ ab^2 = 17ab^2$

24. C
$x^2 - 7x - 30 = 0$ ... We try to separate the middle term -7x to find common factors with x^2 and -30 separately:

$x^2 - 10x + 3x - 30 = 0$ ... Here, we see that x is a common factor for x^2 and -10x, and 3 is a common factor for 3x and -30:

$x(x - 10) + 3(x - 10) = 0$ ... Here, we have x times x - 10 and 3 times x - 10 summed up. This means that we have x + 3 times x - 10:
$(x + 3)(x - 10) = 0$ or $(x - 10)(x + 3) = 0$

25. A
We need to simplify the equation by distributing factors and then collecting x terms on one side, and the others on the other side:

$(a + 2)x - b = -2 + (a + b)x$

$ax + 2x - b = -2 + ax + bx$

$ax + 2x - ax - bx = -2 + b$ ... ax and -ax cancel each other:

$2x - bx = -2 + b$ ... we take -1 as a factor on the right side:

$(2 - b)x = -(2 - b)$

$x = -(2 - b)/(2 - b)$ ... Simplifying by 2 - b:

$x = -1$

26. C
To obtain a polynomial, remove the parenthesis by distributing the related factors to the terms inside the parenthesis:
$1 - x(1 - x(1 - x)) = 1 - x(1 - (x - x * x)) = 1 - x(1 - x + x^2)$

$= 1 - (x - x * x + x * x^2) = 1 - x + x^2 - x^3$ … Writing this result in descending order of powers:

$= - x^3 + x^2 - x + 1$

27. D

To simplify the expression, remove the parenthesis by distributing the related factors to the terms inside the parenthesis:

$7(2y + 8) + 1 – 4(y + 5)$
$= (7 * 2y + 7 * 8) + 1 - (4 * y + 4 * 5)$

$= 14y + 56 + 1 - 4y - 20$

$= 14y - 4y + 56 + 1 - 20$ … similar terms written together to ease summing/substituting.

$= 10y + 37$

28. D

We are given that each of the n employees earns s amount of salary weekly. This means that one employee earns s salary weekly. So; Richard has 'ns' amount of money to employ n employees for a week.

We are asked to find the number of days n employees can be employed with x amount of money. We can do simple direct proportion:

If Richard can employ 'n' employees for 7 days with 'ns' amount of money,

Richard can employ n employees for y days with x amount of money … y is the number of days we need to find.

Cross multiply:

$y = (x * 7)/(ns)$

$y = 7x/ns$

29. A
$x^2 - 3x - 4$ ... try to separate the middle term $-3x$ to find common factors with x^2 and -4 separately:

$x^2 + x - 4x - 4$ ... Here, x is a common factor for x^2 and x, and -4 is a common factor for -4x and -4:

$= x(x + 1) - 4(x + 1)$ ... Here, x times x + 1 and -4 times x + 1 summed up. This means that we have x - 4 times x + 1:

$= (x - 4)(x + 1)$ or $(x + 1)(x - 4)$

30. D
To solve the equation, we need the equation in the form $ax^2 + bx + c = 0$.

$(a^2 - b^2)x^2 + 2ax + 1 = 0$ is already in this form.

The quadratic formula to find the roots of a quadratic equation is:

$x_{1,2} = (-b \pm \sqrt{\Delta}) / 2a$ where $\Delta = b^2 - 4ac$ and is called the discriminant of the quadratic equation.

In our question, the equation is $(a^2 - b^2)x^2 + 2ax + 1 = 0$.

By remembering the form $ax^2 + bx + c = 0$: $a = a^2 - b^2$, $b = 2a$, $c = 1$

So, we can find the discriminant first, and then the roots of the equation:

$\Delta = b^2 - 4ac = (2a)^2 - 4(a^2 - b^2) * 1 = 4a^2 - 4a^2 + 4b^2 = 4b^2$

$x_{1,2} = (-b \pm \sqrt{\Delta}) / 2a = (-2a \pm \sqrt{4b^2}) / (2(a^2 - b^2)) = (-2a \pm 2b)$

$/ (2(a^2 - b^2))$

$= 2(-a \pm b) / (2(a^2 - b^2))$ ... We can simplify by 2:

$= (-a \pm b) / (a^2 - b^2)$

This means that the roots are,

$x_1 = (-a - b) / (a^2 - b^2)$ ... $a^2 - b^2$ is two square differences:

$x_1 = -(a + b) / ((a - b)(a + b))$ ... (a + b) terms cancel:

$x_1 = -1/(a - b)$

$x_2 = (-a + b) / (a^2 - b^2)$ ... $a^2 - b^2$ is two square differences:

$x_2 = -(a - b) / ((a - b)(a + b))$ ... (a - b) terms cancel:

$x_2 = -1/(a + b)$

31. A
To simplify, remove the parenthesis and see if any terms cancel:

$(a + b)(x + y) + (a - b)(x - y) - (ax + by) = ax + ay + bx + by + ax - ay - bx + by - ax - by$

Writing similar terms together:

$= ax + ax - ax + bx - bx + ay - ay + by + by - by$ ... + terms cancel - terms:

$= ax + by$

32. D
The area of a rectangle is found by multiplying the width to the length. If we call these sides with "a" and "b"; the area is = a * b.

We are given that a * b = 20 cm^2 ... Equation I

One side is increased by 1 and the other by 2 cm. So new side lengths are "a + 1" and "b + 2."

The new area is $(a + 1)(b + 2) = 35\ cm^2$ … Equation II

Using equations I and II, we can find a and b:

$ab = 20$

$(a + 1)(b + 2) = 35$ … distribute the terms in parenthesis:

$ab + 2a + b + 2 = 35$

Insert $ab = 20$ to the above equation:

$20 + 2a + b + 2 = 35$

$2a + b = 35 - 2 - 20$

$2a + b = 13$ … This is one equation with two unknowns. We need to use another information to have two equations with two unknowns which leads us to the solution. We know that $ab = 20$. So, we can use $a = 20/b$:

$2(20/b) + b = 13$

$40/b + b = 13$ … equate all denominators to "b" and eliminate it:

$40 + b^2 = 13b$

$b^2 - 13b + 40 = 0$ … use the roots by factoring. We try to separate the middle term -13b to find common factors with b^2 and 40 separately:

$b^2 - 8b - 5b + 40 = 0$ … Here, b is a common factor for b^2 and -8b, and -5 is a common factor for -5b and 40:

$b(b - 8) - 5(b - 8) = 0$ Here, b times b - 8 and -5 times b - 8 summed up. This means that we have b - 5 times b - 8:

$(b - 5)(b - 8) = 0$

This is true when either or both of the expressions in the parenthesis are equal to zero:

b - 5 = 0 ... b = 5

b - 8 = 0 ... b = 8

So we have two values for b which means we have two values for a as well. To find a, we can use any equation we have. Let us use a = 20/b.

If b = 5, a = 20/b → a = 4

If b = 8, a = 20/b → a = 2.5

So, (a, b) pairs for the sides of the original rectangle are: (4, 5) and (2.5, 8). These are found in (b) and (c) answer choices.

33. B

Finding the x-intercepts of a function means that we need to equate the function to zero and find the roots of the equation:

$(x - 5)^2 - 9 = 9$

$(x - 5)^2 = 9$

$\sqrt{(x - 5)^2} = \sqrt{9}$

x - 5 = 3 → x = 8

x - 5 = -3 → x = 2

34. B

We have two variables: the price of a t-shirt and a pair of pants; and we have two situations given about them. We need to set two equations and solve them for the variables. Then, we are asked to find the difference.
Let us call the price of a t-shirt by a, and the price of a

pair of pants by b:
If John buys 4 t-shirts and 5 pair of pants, he pays $51 → 4a + 5b = 51

If he buys 7 t-shirts and 3 pair of pants, then he pays $49 → 7a + 3b = 49

4a + 5b = 51

7a + 3b = 49

We have two paths to follow: substitution or elimination. Here, since extracting a or b from either equation results in fractions; it is easier to choose elimination:

-3/ 4a + 5b = 51

5/ 7a + 3b = 49

-12a - 15b = -153

35a + 15b = 245

23a = 92

a = 4

Choosing either of the equations, find b, by inserting a:

4 * 4 + 5b = 51

16 + 5b = 51

5b = 35

b = 7

The difference between a and b is 7 - 4 = 3.

35. D
The simplest way to draw the graph of a linear equation is to insert zero into x and y separately and to obtain two points on the line.

$4x - y = 6$ is the equation of the line.

If $x = 0$, $y = -6 \rightarrow$ point $(0, -6)$ is obtained

If $y = 0$, $x = 6/4 = 1.5 \rightarrow$ point $(1.5, 0)$ is obtained

Line $4x - y = 6$ passes through $(0, -6)$ and $(1.5, 0)$

The graph satisfying the condition is given in choice D.

36. B
Let us call this number by x:

This number is increased by 2: $x + 2$

Then, it is multiplied by 3: $3(x + 2)$

The result is 24: $3(x + 2) = 24$ … Solving this linear equation, we obtain the value of the number:

$x + 2 = 24 / 3$
$x + 2 = 8$
$x = 8 - 2$
$x = 6$

37. B
My age: x

My brother is 3 years older than me: $x + 3$

My father is 3 less than 2 times my age: $2x - 3$

My father's age divided by 5 is equal to my brother's age

divided by 3: $(2x - 3) / 5 = (x + 3) / 3$
By cross multiplication: $5(x + 3) = 3(2x - 3)$

$5x + 15 = 6x - 9$
$x = 24$

My father's age: 2.24 – 3= 48 – 3 = 45

38. C
There are two fractions containing x and the denominators are different. First, find a common denominator to simplify the expression. The least common multiplier of 4 and 7 is 28. Then,

$7(x - 2) / 28 - 4(3x + 5) / 28 = -3.28 / 28$ ... Since both sides are written on the denominator 28 now, we can eliminate them:

$7(x - 2) - 4(3x + 5) = -84$

$7x - 14 - 12x - 20 = -84$
$-5x = - 84 + 14 + 20$
$-5x = -50$
$x = 50/5$
$x = 10$

39. C
The inner angles of a triangle sum up to 180°. Let us sum three expressions given for the inner angles equating to 180° and then find x:

$(x + 20) + (3x - 10) + (8x + 50) = 180$
$x + 3x + 8x + 20 - 10 + 50 = 180$
$12x + 60 = 180$
$12x = 120$
$x = 10$

Without calculation, it is obvious that 8x + 50 is the largest angle, but we cannot know which of the remaining two expressions gives the smallest value; so calculate each:

$x + 20 = 10 + 20 = 30$
$3x - 10 = 30 - 10 = 20$
$8x + 50 = 80 + 50 = 130$

The largest angle is 130° and the smallest is 20°. Their difference is 130 – 20 = 110°.

40. B
The width of the rectangle is given to be two thirds of the length. So, if we call the length by a, the width should be (2/3)a. To deal with fractions, let us say that:

length = 3x

Then, width = (2/3)3x = 2x

Remember that the perimeter of a rectangle is found by summing all sides, which means summing two lengths and two widths.

Perimeter = 2.3x + 2.2x = 150
$6x + 4x = 150$
$10x = 150$
$x = 15$

We are asked to find the length, that is
3x = 3.15 = 45 cm.

41. D
The volume of a sphere is found by $V_{sphere} = (4/3)\pi r^3$

The volume of a cylinder is found by $V = \pi r^2 h$ ... In the question, the radius of the cylinder is 3r. Then,

$V_{cylinder} = \pi(3r)^2h = 9\pi r^2h$

Since in this question $V_{sphere} = V_{cylinder}$,

$(4/3)\pi r^3 = 9\pi r^2h$ … Eliminating πr^2 from both sides:

$(4/3)r = 9h$ … By cross multiplication:
$4r = 27h$

We are asked to find r/h. From the above equation, we can say that if r = 27k, then h = 4k.

r/h = 27k / 4k = 27/4.

42. A
The area of a triangle is found by the formula
Area = (1/2) * base *height

Let us say that base = x
The height of this triangle is 4 cm less than 3 times the base; so, height = 3x – 4

Applying these to the equation above:

Area = 32 = (1/2) * x * (3x – 4)

By cross multiplication and distributing the parenthesis,

$64 = 3x^2 - 4x$
$3x^2 - 4x - 64 = 0$ … By factorization,

3x –16

x 4
$(3x - 16)(x + 4) = 0$
There are two solutions for x:
1) $3x - 16 = 0 \rightarrow x = 16/3$

2) $x + 4 = 0 \rightarrow x = -4$ … Since a length measure cannot be negative, this cannot be a possible solution.

The only solution for x is 16/3.

We are asked to find the height, that is $3x - 4 = 3(16/3) - 4 = 16 - 4 = 12$ cm.

43. C

Since we do not know the value of x, we do not know if $5x - 3$ is negative or positive. Therefore, we need to consider all possible solutions.

$5x - 3 \leq 7$ and $-(5x - 3) \leq 7$ are two possible cases. Let us organize the second one:

$-(5x - 3) \leq 7$ … Multiply both sides by -1 and do not forget to change the direction of the inequality:

$5x - 3 \geq -7$

Now, we can combine both cases for $|5x - 3| \leq 7$:
$-7 \leq 5x - 3 \leq 7$

We are searching for a solution for x alone. Therefore, let us add 3 to all sides:

$-7 + 3 \leq 5x - 3 + 3 \leq 7 + 3$
$-4 \leq 5x \leq 10$

Now, let us divide all sides by 5:
$-4/5 \leq x \leq 2$

Integer values within this interval are 0, 1 and 2. The sum of these is $0 + 1 + 2 = 3$.

44. A

Since we do not know the value of x, we do not know if $7 - 2x$ is negative or positive. Therefore, we need to consider all possible solutions.

If $7 - 2x$ is positive, then $|7 - 2x| = 7 - 2x \rightarrow 7 - 2x = 9$
$2x = -2 \rightarrow x = -1$

If $7 - 2x$ is negative, then $|7 - 2x| = -(7 - 2x) = 2x - 7 \rightarrow$
$2x - 7 = 9$

$2x = 16 \rightarrow x = 8$

Multiplying the two solutions: $(-1)(8) = -8$.

45. A

First let us leave the absolute value alone by adding -8 to both sides:

$|x - 5| + 8 - 8 < 4 - 8$

$|x - 5| < -4$

This inequality tells that the result of the absolute value is negative and is smaller than -4. However, the result of an absolute value can be zero the smallest; it cannot be a negative number. Then in this question, there is no solution for the inequality given.

Basic Math Multiple Choice

How to Answer Basic Math Questions - the Basics

First, read the problem, but not the answers.

Work through the problem first and come up with your own answers. Hopefully, you should find your answer among the choices.

If no answer matches the one you got, re-check your math, but this time, use a different method. In math, there are different ways to solve a problem.

Math Multiple Choice Strategy

The two strategies for working with basic math multiple choice are Estimation and Elimination.

Estimation is just as it sounds - try to estimate an approximate answer first. Then look at the choices.

Elimination is probably the most powerful strategy for answering multiple choice.

Eliminate obviously incorrect answers and narrowing the possible choices.

Here are a few basic math examples of how this works.

Solve 2/3 + 5/12

a. 9/17
b. 3/11
c. 7/12
d. 1 1/12

First estimate the answer. 2/3 is more than half and 5/12 is about half, so the answer is going to be very close to 1.

Next, Eliminate. Choice A is about 1/2 and can be eliminated, choice B is very small, less than 1/2 and can be eliminated. Choice C is close to 1/2 and can be eliminated. Leaving only choice D, which is just over 1.

Work through the solution, find a common denominator and add. The correct answer is 1 1/12, so Choice D is correct.

Let's look at another example:

Solve 4/5 – 2/3

a. 2/2
b. 2/13
c. 1
d. 2/15

First, quickly estimate the answer. 4/5 is very close to 1, and 2/3 more than half, so the answer is going to be less than 1/2.

Choice A can be eliminated right away, because it is 1. Choice C can be eliminated for the same reason.

Next, look at the denominators. Since 5 and 3 don't go into 13, choice B can be eliminated as well. That leaves choice D. Checking the answer, the common denominator will be 15. So the answer is 2/15 and choice D is correct.

Fractions Shortcut - Canceling Out

In any operation with fractions, if the numerator of one fraction has a common multiple with the denominator of the other, you can cancel out. This saves time and simplifies the problem quickly, making it easier to manage.

Solve 2/15 ÷ 4/5

a. 6/65

b. 6/75

c. 5/12

d. 1/6

To divide fractions, we multiply the first fraction with the inverse of the second fraction. Therefore we have 2/15 x 5/4. The numerator of the first fraction, 2, shares a multiple with the denominator of the second fraction, 4, which is 2. These cancel out, which gives, 1/3 x 1/2 = 1/6

Canceling out solved the questions very quickly, but we can still use multiple choice strategies to answer.

Choice B can be eliminated because 75 is too large a denominator. Choice C can be eliminated because 5 and 15 don't go into 12.

Choice D is correct.

Basic Math Multiple Choice Strategy and Short-cuts

Multiplying decimals gives a very quick way to estimate and eliminate choices. Anytime that you multiply decimals, it is going to give an answer with the same number of decimal places as the combined operands.

So for example,

2.38 X 1.2 will produce a number with three places of decimal, which is 2.856.
Here are a few examples with step-by-step explanation:

Solve 2.06 x 1.2

a. 24.82
b. 2.482
c. 24.72
d. 2.472

This is a simple question, but even before you start calculating, you can eliminate several choices. When multiplying decimals, there will always be as many numbers behind the decimal place in the answer as the sum of the ones in the initial problem, so choices A and C can be eliminated.

The correct answer is D: 2.06 x 1.2 = 2.472

Solve 20.0 ÷ 2.5

a. 12.05

b. 9.25

c. 8.3

d. 8

First estimate the answer to be around 10, and eliminate choice A. And since it'd also be an even number, you can eliminate choices B and C, leaving only choice D.

The correct Answer is D: 20.0 ÷ 2.5 = 8

How to Study for a Math Test

EVERY SUBJECT HAS ITS OWN PARTICULAR STUDY METHOD. Math is mostly numerical, not verbal and requires logical thinking; it has its own way to be studied. Before touching on significant points of studying a math test, lets look at some of the fundamentals of "learning."

Learning is not an instant experience; it is a procedure. Learning is a process not an event. Rome wasn't built in a day, and learning anything (or everything) isn't going to happen in a day either. You cannot expect to learn everything in one day, at night, before the test. It is important and necessary to learn day-by-day. Good time management plays a considerable role in learning. When you manage your time, and begin test preparation well in advance, you will notice the subjects are easier than you thought, or feared, and you will take the test without the stress of a sleepless body and an anxious mind.

Memorizing is a temporary step of learning if information is not comprehended and applied afterwards. Memorize just the basics and understand the meaning; then apply, analyze, synthesize and evaluate.

These are the hierarchical layout of cognitive learning: Of course, there are some basic properties that you need to memorize in the beginning, since you cannot prove the facts every time you solve a math test. For example; the inner angles of a triangle sum up to 180°. If you do not know this, you may not be able to solve triangle problems. And, more importantly, if you do not practice, you will certainly forget this property. Practice helps information take root in your brain. Applying the same property on various types of questions extends the roots.

For example, if you see a triangle, you can analyze the question by means of the property. In a question, if you see a hexagon, you can split it into triangles and use the property, called synthesizing followed by evaluation. If all these steps are followed, the property is completely learned and has its place in your long term memory.

A useful method in providing consistent learning is using similarities between the information and events, images, shapes, ... etc. For example, assume that you have difficulty remembering the formula $x^a/y^b = x^a y^{-b} = 1/x^{-a}y^b$ in mind. You can associate this to an elevator: The exponents changing location (nominator/denominator) need to change exponent sign, similarly, people going up need to push the up button and if they decide to go down, they need to push the down button; so they need to change the button. Also; writing the formula in large letters and sticking it on a surface that is frequently visible helps memorizing it by using visual intelligence. The more senses (visual, musical, auditory, logical, ...) the material addresses to, the more permanent it is.

Attend to all classes. Knowledge is not replaceable by others, and every brain is unique. You cannot learn math from your classmate's notes; take your own notes in your own understanding of the material. What you understand, or don't understand, and how you understand it is different to everyone else. Highlight the important points in your own way. Remember math and all other courses are mostly learned at school -practice comes afterwards at home.

Find your own way of learning. Every person learns and studies differently. Some take notes, some do not like writing; listening is the major way of embracing information for them, and some watch. It is important to detect the way that is more useful for you. Coloring important points also helps. Due to selective perception; we see the attractive words, signs before the rest. While studying math, make a list; first, determine the subjects you feel inadequate on and focus on them initially.

Never gloss over something that you do not fully understand. Information is built on previous learning in a hierarchical order. If you have questions about a mathematical property, and don't understand it completely, you cannot solve problems using that property. You need to have a strong background to succeed in math. You need to know your basic math inside out to do algebra. And you need to know your algebra inside out to do calculus. If you do not know exponentials, you cannot solve logarithms. If you don't understand something, get help from your teachers, reread course materials, resolve examples, discuss with friends or hire a private tutor. Never skip over something that you don't understand – it will come back to haunt you!

Practice makes perfect! Yes – it really does! Working through math problems in your own way is essential. Looking over examples is a good first step – but only that. The example solutions shown in the textbook show you have to solve a problem, next you have to do it yourself. Do not miss to do your assignments. You see different types of examples and acquire different outlooks when facing math problems. Math is fun because usually there are many ways to reach the solution. Find alternative ways to solve a problem which anchors the learning deeper. Find similar and different problems, discuss with friends, ask each other questions.

Observing other people's way of thinking, and solving problems will help both of you improves your minds.

Succeeding in math is a mental action. However, do not disregard physical and psychological effects. Always think positive and never give up; no success is gained without effort. Of course you will waste time solving problems which you will find easy, and you will struggle with difficult problems. In the end, you will be one step further ahead. And after more time spent practicing, you will be another step ahead.

Reward yourself after intense studies. This will keep your motivation high. The reward may be a chocolate bar, playing a game for 20 minutes, or taking a walk in the park. It is very essential to have a good night's sleep before taking a math test. Eating habits directly effect success. Keep away from fast food as much as you can, eat a light meal before a test. At last but not least, keep your inner motivation very high. Believe in yourself; you will certainly get the good result of your planned, efficient studies.

Quick Tips - Studying for a Math Test

1. Use Frequent Practice & Quizzes (Testing Effect)

Tip: Regular self-testing with practice problems or quizzes helps solidify memory and retention.

Why it helps: Active recall (or retrieval practice) reinforces long-term learning far more effectively than passive review.

2. Break Study Sessions into Spaced, Short Blocks

Tip: Study in short sessions (e.g., 25–30 minutes) with regular breaks

Why it helps: Spaced sessions prevent burnout, increase focus, and align with efficient learning rhythms.

3. Master the Calculator & Formula Sheet Ahead of Time

Tip: Familiarize yourself with the GED's on-screen calculator (TI-30XS) and reference sheet during study..

Why it helps: Building familiarity increases confidence during the test.

4. Identify Weak Areas with Diagnostic Practice Tests

Tip: Take a full practice test to spot weak areas (e.g., algebra, geometry), then focus your study accordingly.

Why it helps: Targeted practice builds confidence and ensures efficient use of study time.
Practice Questions
https://test-preparation.ca/practice-questions-for-standardized-tests/

5. Study with Friends

Tip: Join or form study groups to work through challenging problems together.

Why it helps: Explaining concepts to others and hearing different approaches deepens your own understanding.
https://test-preparation.ca/learning-study/#group

6. Over-learn & Review for Long-Term Confidence

Tip: Don't stop after understanding—revisit topics multiple times until they become second nature.

Why it helps: Over-learning solidifies knowledge and supports recall under stress.

7. Eliminate Distractions & Practice Mindful Study

Tip: Find a quiet study space, turn off notifications, and focus on the material at hand.

Why it helps: Reducing noise and interruptions boosts concentration and cognitive efficiency.
https://test-preparation.ca/learning-study/

8. Use Mnemonics & Memory Techniques

Tip: Employ memory aids like the method of loci or simple acronyms to help memorize formulas or steps.

Why it helps: These techniques improve recall and make study sessions more engaging.

https://test-preparation.ca/a-guide-to-memorizing-anything-easily-and-painlessly/

How to Prepare for a Test

MOST STUDENTS HIDE THEIR HEADS AND PROCRASTINATE WHEN FACED WITH PREPARING FOR AN EXAM, HOPING THAT SOMEHOW THEY WILL BE SPARED THE AGONY, ESPECIALLY IF IT IS A BIG ONE THAT THEIR FUTURES RELY ON. Avoiding a test is what many students do best and unfortunately, they suffer the consequences because of their lack of preparation.

Preparing for a test involves building your knowledge base and refining skills—studying concepts, practicing under timed conditions, simulating the test format, and building confidence. It's about planning, engaging deeply with material, and building mental readiness. You can rehearse, pause, and correct mistakes during your prep.

Test preparation requires strategy and dedication. It is the perfect training ground for a professional life. Besides having several reliable strategies, successful students also has a clear goal and know how to accomplish it. These tried and true concepts have worked well and will make your test preparation easier.

Test Prep and Study Skills Video Tutorials

https://www.test-preparation.ca/test-video/

Test Prep - the Basics

Take responsibility for your own test preparation.

It is a common - but big - mistake to link your studying to someone else's. Study partners are great, but only if they are reliable. It is your job to be prepared for the test, even if a study partner fails you. Do not allow others to distract you from your goals.

Prioritize the time available to study

When do you learn best, early in the day or at night? Does your mind absorb and retain information most efficiently in small blocks of time, or do you require long stretches to get the most done? It is important to figure out the best blocks of time available to you when you can be the most productive. Try to consolidate activities to allow for longer periods of study time.

Find a quiet place where you will not be disturbed

Do not try to squeeze in quality study time in any old location. Find a peaceful place with a minimum of distractions, such as the library, a park or even the laundry room. Good lighting is essential and you need to have comfortable seating and a desk surface large enough to hold your materials. It is probably not a great idea to study in your bedroom. You might be distracted by clothes on the floor, a book you have been planning to read, the telephone or something else. Besides, in the middle of studying, that bed will start to look very comfortable. Whatever you do, avoid using the bed as a place to study since you might fall asleep to avoiding studying!

The exception is flashcards. By far the most productive study time is sitting down and studying and studying only. However, with flashcards you can carry them with you and make use of odd moments, like standing in line or waiting for the bus. This isn't as productive, but it really helps and is definitely worth doing.

Determine what you need to study

Gather together your books, your notes, your laptop and any other materials needed to focus on your study for this exam. Ensure you have everything you need so you don't waste time. Remember paper, pencils and erasers, sticky notes, bottled water and a snack. Keep your phone with you if you need it to find essential information, but keep it turned off so others can't distract you.

Have a positive attitude

It is essential that you approach your studies for the test with an attitude that says you will pass it. And pass it with flying colors! This is one of the most important keys to successful studying. Believing that you are capable helps you to become capable.

The Strategy of Studying

Review class notes

Stay on top of class notes and assignments by reviewing them frequently and regularly. Re-writing notes can be a terrific study trick, as it helps lock in information. Pay special attention to any comments that have been made by the teacher. If a study guide has been made available as part of the class materials, use it! It will be a valuable tool to use for studying.

Estimate how much time you will need

If you are concerned about the amount of time you have available it is a good idea to set up a schedule so that you do not get bogged down on one section and end without enough time left to study other things. Remember to schedule break time, and use that time for a little exercise or other stress reducing techniques.

Test yourself to determine your weaknesses

Look online for additional assessment and evaluation tools available for a particular subject. Visit our website https://www.test-preparation.ca for test tips and more practice questions. Once you have determined areas of concern, you will be able to focus on studying the information they contain and just brush up on the other areas of the exam.

Mental Prep – How to Psych Yourself Up for a Test

Since tests are often a big factor in your final grade or acceptance into a program, it is understandable that taking tests creates anxiety for many students. Even students who know they have learned the required material find their minds going blank as they stare at the test booklet. One easy way to overcome that anxiety is to prepare mentally for the test. Here are a few simple techniques.

Do not procrastinate

Study the material for the test when it becomes available, and continue to review the material until the test day. By waiting until the last minute and trying to cram for the test the night before, you actually increase anxiety. This leads to an increase in negative self-talk. Telling yourself "I can't learn this. I am going to fail" is a pretty sure indication that you are right. At best, your performance on the test will not be as strong if you have procrastinated instead of studying.

Positive self-talk.

Positive self-talk drowns out negative self-talk and to increases your confidence level. Whenever you begin feeling overwhelmed or anxious about the test, remind yourself that you have studied enough, you know the material and that you will pass the test. Both negative and positive self-talk are really just your fantasy, so why not choose to be a winner?

Do not compare yourself to others.

Do not compare yourself to other students. Instead, focus on your strengths and weaknesses and prepare accordingly. Regardless of how others perform, your performance is the only one that matters to your grade. Comparing yourself to others increases your anxiety and negative self-talk before the test.

Visualize.

Make a mental image of yourself taking the test. You know the answers and feel relaxed. Visualize doing well on the test and having no problems with the material. Visualizations can increase your confidence and decrease the anxiety you might otherwise feel before the test. Instead of thinking of this as a test, see it as an opportunity to demonstrate what you have learned!

Avoid negativity.

Worry is contagious and viral - once it gets started it builds on itself. Cut it off before it gets to be a problem. Even if you are relaxed and confident, being around anxious, worried classmates might cause you to start feeling anxious. Before the test, tune out the fears of classmates. Feeling anxious and worried before an exam is normal, and every student experiences those feelings at some point. But you cannot allow these feelings to interfere with your ability to perform well. Practicing mental preparation techniques and remembering that the test is not the only measure of your academic performance will ease your anxiety and ensure that you perform at your best.

Complete Guide to Test Preparation

https://test-preparation.ca/the-ultimate-guide-to-test-preparation-strategy/

How to Take a Test

EVERYONE KNOWS THAT TAKING AN EXAM IS STRESSFUL, BUT IT DOES NOT HAVE TO BE THAT BAD! There are a few simple things that you can do to increase your score on any type of test. Take a look at these tips and consider how you can incorporate them into your study time.

OK - so you are in the test room - Here is what to do!

Video Tutorials
https://www.test-preparation.ca/test-video/

Reading the Instructions

This is the most basic point, but one that, surprisingly, many students ignore and it costs big time! Since reading the instructions is one of the most common, and 100% preventable mistakes, we have a whole section just on reading instructions.

Pay close attention to the sample questions. Almost all standardized tests offer sample questions, paired with their correct solutions. Go through these to make sure that you understand what they mean and how they arrived at the correct answer. Do not be afraid to ask the test supervisor for help with a sample that confuses you, or instructions that you are unsure of.

Tips for Reading the Question

We could write pages and pages of tips just on reading the test questions. Here are a few that will help you the most.

- **Think first.** Before you look at the answer, read and think about the question. It is best to try to come up with the correct answer before you look at the options. This way, when the test-writer tries to trick you with a close answer, you will not fall for it.

- **Make it true or false.** If a question confuses you, then look at each answer option and think of it as a "true" "false" question. Select the one that seems most likely to be "true."

- **Mark the Question.** For some reason, a lot of test-takers are afraid to mark up their test booklet. Unless you are specifically told not to mark in the booklet, you should feel free to use it to your advantage.

- **Circle Key Words.** As you are reading the question, underline or circle key words. This helps you to focus on the most critical information needed to solve the problem. For example, if the question said, "Which of these is not a synonym for huge?" You might circle "not," "synonym" and "huge." That clears away the clutter and lets you focus on what is important.

- **Always underline these words:** all, none, always, never, most, best, true, false and except.

- **Eliminate.** Elimination is the best strategy for multiple choice answers *and* questions. If you are confused by lengthy questions, cross out anything that you think is irrelevant, obviously wrong, or information that you think is offered to distract you.

- **Do not try to read between the lines.** Usually, questions are written to be straightforward, with no deep, underlying meaning. Generally, the simple answer really is the correct answer. Do not over-analyze!

How to Take a Test - The Basics

Some sections of the test are designed to assess your ability to quickly grab the necessary information; this type of exam makes speed a priority. Others are more concerned with your depth of knowledge, and how accurate it is. When you start a new section of the test, look it over to determine whether the test is for speed or accuracy. If the test is for speed (a lot of questions and a short time), your strategy is clear; answer as many questions as quickly as possible.

The HiSET© does NOT penalize for wrong answers, so if all else fails, guess and make sure you answer every question.

Every little bit helps

For information on the GED® calculator policy see,
http://www.gedtestingservice.com/testers/calculator
Make sure you know how to use it!

You cannot bring any other materials into the test room. Scratch paper and a pencil are provided. Use them!

Make time your friend

Budget your time from the beginning until you are finished, and stick to it! The amount of time you are permitted for each portion of the test will almost certainly be included in the instructions.

Easy does it

One smart way to tackle a test is to locate the easy questions and answer those first. This is a time-tested strategy that never fails, because it saves you a lot of unnecessary anxiety. First, read the question and decide if you can answer it in less than a minute. If so, complete the question and go to the next one. If not, skip it for now and continue to the next question. By the time you have completed the first pass through this section of the exam, you will have answered a good number of questions. Not only does it boost your confidence, relieve anxiety and kick your memory up a notch, you will know exactly how many questions remain and can allot the rest of your time accordingly. Think of doing the easy questions first as a warm-up!

Do not watch your watch

At best, taking an important exam is an uncomfortable situation. If you are like most people, you might be tempted to subconsciously distract yourself from the task at hand. One of the most common ways to do so is by becoming obsessed with your watch or the wall clock. Do not watch your watch! Take it off and place it on the top corner of your desk, far enough away that you will not be tempted to look at it every two minutes. Better still, turn the watch face away from you. That way, every time you try to sneak a peek, you will be reminded to refocus your attention to the task at hand. Give yourself permission to check your watch or the wall clock after you complete each section. Focus on answering the questions, not on how many minutes have elapsed since you last looked at it.

Divide and conquer

What should you do when you come across a question that is so complicated you may not even be certain what is being asked? As we have suggested, the first time through, skip the question. At some point, you will need to return to it and get it under control. The best way to handle questions that leave you feeling so anxious you can hardly think is by breaking them into manageable pieces. Solving smaller bits is always easier. For complicated questions, divide them into bite-sized pieces and solve these smaller sets separately. Once you understand what the reduced sections are really saying, it will be much easier to put them together and get a handle on the bigger question. This may not work with every question - see below for how to deal with questions you cannot break down.

Reason your way through the toughest questions

If you find that a question is so dense you can't figure out how to break it into smaller pieces, there are a few strategies that might help. First, read the question again and look for hints. Can you re-word the question in one or more different ways? This may give you clues. Look for words that can function as either verbs or nouns, and try to figure out what the questions is asking from the sentence structure. Remember that many nouns in English have several different meanings. While some of those meanings might be related, sometimes they are completely distinct. If reading the sentence one way does not make sense, consider a different definition or meaning for a key word.

The truth is, it is not always necessary to understand a question to arrive at a correct answer! The most successful strategy for multiple choice is Elimination. Frequently, at least one answer is clearly wrong and can be crossed off the list of possible correct answers. Next, look at the remaining answers and eliminate any that are only partially true. You may still have to flat-out guess from time to time, but using the process of elimination will help you make your way to the correct answer more often than not - even when you don't know what the question means!

Do not leave early

Use all the time allotted to you, even if you can't wait to get out of the testing room. Instead, once you have finished, spend the remaining time reviewing your answers. Go back to those questions that were most difficult for you and review your response. Another good way to use this time is to return to multiple-choice questions in which you filled in a bubble. Do a spot check, reviewing every fifth or sixth question to make sure your answer coincides with the bubble you filled in. This is a great way to catch yourself if you made a mistake, skipped a bubble and therefore put all your answers in the wrong bubbles!

Become a super sleuth and look for careless errors. Look for questions that have double negatives or other odd phrasing; they might be an attempt to throw you off. Careless errors on your part might be the result of skimming a question and missing a key word. Words such as "always", "never", "sometimes" , "rarely" and the like can give a strong indication of the answer the question is really seeking. Don't throw away points by being careless!

Just as you budgeted time at the beginning of the test to allow for easy and more difficult questions, be sure to budget sufficient time to review your answers. On essay questions and math questions where you are required to show your work, check your writing to make sure it is legible.

Math questions can be especially tricky. The best way to double check math questions is by figuring the answer using a different method, if possible.

Here is another terrific tip. It is likely that no matter how hard you try, you will have a handful of questions you just are not sure of. Keep them in mind as you read through the rest of the test. If you can't answer a question, looking back over the test to find a different question that addresses the same topic might give you clues.

We know that taking the test has been stressful and you can hardly wait to escape. Just keep in mind that leaving before you double-check as much as possible can be a quick trip to disaster. Taking a few extra minutes can make the difference between getting a bad grade and a great one. Besides, there will be lots of time to relax and celebrate after the test is turned in.

In the Test Room – What you MUST do!

If you are like the rest of the world, there is almost nothing you would rather avoid than taking a test. Unfortunately, that is not an option if you want to pass. Rather than suffer, consider a few attitude adjustments that might turn the experience from a horrible one to...well, an interesting one! Take a look at these tips. Simply changing how you perceive the experience can change the experience itself.

You have to take the test - you can't change that. What you can change, and the only thing that you can change, is your attitude -so get a grip - you can do it!

Get in the mood

After weeks of studying, the big day has finally arrived. The worst thing you can do to yourself is arrive at the test site feeling frustrated, worried, and anxious. Keep a check on your emotional state. If your emotions are shaky before a test it can determine how well you do on the test. It is extremely important that you pump yourself up, believe in yourself, and use that confidence to get in the mood!

Don't fight reality

Students often resent tests, and with good reason. After all, many people do not test well, and they know the grade they end with does not accurately reflect their true knowledge. It is easy to feel resentful because tests classify students and create categories that just don't seem fair. Face it: Students who are great at rote memorization and not that good at actually analyzing material often score higher than those who might be more creative thinkers and balk at simply memorizing cold, hard facts. It may not be fair, but there it is anyway. Conformity is an asset on tests, and creativity is often a liability. There is no point in wasting time or energy being upset about this reality. Your first step is to accept the reality and get used to it. You will get higher marks when you realize tests do count and that you must give them your best effort. Think about your future and the career that is easier to achieve if you have consistently earned high grades. Avoid negative energy and focus on anything that lifts your enthusiasm and increases your motivation.

Get there early enough to relax

If you are wound up, tense, scared, anxious, or feeling rushed, it will cost you. Get to the exam room early and relax before you go in. This way, when the exam starts, you are comfortable and ready to apply yourself. Of course, you do not want to arrive so early that you are the only one there. That will not help you relax; it will only give you too much time to sit there, worry and get wound up all over again.

If you can, visit the room where you will be taking your exam a few days ahead of time. Having a visual image of the room can be surprisingly calming, because it takes away one of the big 'unknowns'. Not only that, but once you have visited, you know how to get there and will not be worried about getting lost. Furthermore, driving to the test site once lets you know how much time you need to allow for the trip. That means three potential stressors have been eliminated all at once.

Get it down on paper

One advantage of arriving early is that it allows you time to recreate notes. If you spend a lot of time worrying about whether you will be able to remember information like names, dates, places, and mathematical formulas, there is a solution for that. Unless the exam you are taking allows you to use your books and notes, (and very few do) you will have to rely on memory. Arriving early gives to time to tap into your memory and jot down key pieces of information you know will be asked. Just make certain you are allowed to make notes once you are in the testing site; not all locations will permit it. Once you get your test, on a small piece of paper write down everything you are afraid you will forget. It will take a minute or two but by dumping your worries onto the page you have effectively eliminated a certain amount of anxiety and driven off the panic you feel.

Get comfortable in your chair

Here is a clever technique that releases physical stress and helps you get comfortable, even relaxed in your body. You will tense and hold each of your muscles for just a few seconds. The trick is, you must tense them hard for the technique to work. You might want to practice this technique a few times at home; you do not want an unfamiliar technique to add to your stress just before a test, after all! Once you are at the test site, this exercise can always be done in the rest room or another quiet location.

Start with the muscles in your face then work down your body. Tense, squeeze and hold the muscles for a moment or two. Notice the feel of every muscle as you go down your body. Scowl to tense your forehead, pull in your chin to tense your neck. Squeeze your shoulders down to tense your back. Pull in your stomach all the way back to your ribs, make your lower back tight then stretch your fingers. Tense your leg muscles and calves then stretch your feet and your toes. You should be as stiff as a board throughout your entire body.

Now relax your muscles in reverse starting with your toes. Notice how all the muscles feel as you relax them one by one. Once you have released a muscle or set of muscles, allow them to remain relaxed as you proceed up your body. Focus on how you are feeling as all the tension leaves. Start breathing deeply when you get to your chest muscles. By the time you have found your chair, you will be so relaxed it will feel like bliss!

Fight distraction

A lucky few are able to focus deeply when taking an important examination, but most people are easily distracted, probably because they would rather be any place else! There are a number of things you can do to protect yourself from distraction.

Stay away from windows. If you select a seat near a window you may end gazing out at the landscape instead of paying attention to the work at hand. Furthermore, any sign of human activity, from a single individual walking by to a couple having an argument or exchanging a kiss will draw your attention away from your important work. What goes on outside should not be allowed to distract you.

Choose a seat away from the aisle so you do not become distracted by people who leave early. People who leave the exam room early are often the ones who fail. Do not compare your time to theirs.

Of course, you love your friends; that's why they are your friends! In the test room, however, they should become complete strangers inside your mind. Forget they are there. The first step is to physically distance yourself from friends or classmates. That way, you will not be tempted to glance at them to see how they are doing, and there will be no chance of eye contact that could either distract you or even lead to an accusation of cheating. Furthermore, if they are feeling stressed because they did not spend the focused time studying that you did, their anxiety is less likely to permeate your hard-earned calm.

Of course, you will want to choose a seat where there is sufficient light. Nothing is worse than trying to take an important examination under flickering lights or dim bulbs.

Ask the instructor or exam proctor to close the door if there is a lot of noise outside. If the instructor or proctor is unable to do so, block out the noise as best you can. Do not let anything disturb you.

The HiSET© does not allow any personal items in the exam room. Eat protein, complex carbohydrates and a little fat to keep you feeling full and to supercharge your energy. Nothing is worse than a sudden drop in blood sugar during an exam.

Do not allow yourself to become distracted by being too cold or hot. Regardless of the weather outside, carry a sweater, scarf or jacket if the air conditioning at the test site is set too high, or the heat set too low. By the same token, dress in layers so that you are prepared for a range of temperatures.

Watch Caffeine

Drinking a gallon of coffee or gulping a few energy drinks might seem like a great idea, but it is, in fact, a very bad one. Caffeine, pep pills or other artificial sources of energy are more likely to leave you feeling rushed and ragged. Your brain might be clicking along, all right, but chances are good it is not clicking along on the right track! Furthermore, drinking lots of coffee or energy drinks will mean frequent trips to the rest room. This will cut into the time you should be spending answering questions and is a distraction in itself, since each time you need to leave the room you lose focus. Pep pills will only make it harder for you to think straight when solving complicated problems on the exam.

At the same time, if anxiety is your problem try to find ways around using tranquilizers during test-taking time. Even medically prescribed anti-anxiety medication can make you less alert and even decrease your motivation. Being motivated is what you need to get you through an exam. If your anxiety is so bad that it threatens to interfere with your ability to take an exam, speak to your doctor and ask for documentation. Many testing sites will allow non-distracting test rooms, extended testing time and other accommodations as long as a doctor's note that explains the situation is made available.

Keep Breathing

It might not make a lot of sense, but when people become anxious, tense, or scared, their breathing becomes shallow and, in some cases, they stop breathing all together! Pay attention to your emotions, and when you are feeling worried, focus on your breathing. Take a moment to remind yourself to breathe deeply and regularly. Drawing in steady, deep breaths energizes the body. When you continue to breathe deeply you will notice you exhale all the tension.

It is a smart idea to rehearse breathing at home. With continued practice of this relaxation technique, you will begin to know the muscles that tense up under pressure. Call these your "signal muscles." These are the ones that will speak to you first, begging you to relax. Take the time to listen to those muscles and do as they ask. With just a little breathing practice, you will get into the habit of checking yourself regularly and when you realize you are tense, relaxation will become second nature.

Avoid Anxiety Before a Test

Manage your time effectively

This is a key to your success! You need blocks of uninterrupted time to study all the pertinent material. Creating and maintaining a schedule will help keep you on track, and will remind family members and friends that you are not available. Under no circumstances should you change your blocks of study time to accommodate someone else, or cancel a study session to do something more fun. Do not interfere with your study time for any reason!

Relax

Use whatever works best for you to relieve stress. Some folks like a good, calming stretch with yoga, others find expressing themselves through journaling to be useful. Some hit the floor for a series of crunches or planks, and still others take a slow stroll around the garden. Integrate a little relaxation time into your schedule, and treat that time, too, as sacred.

Eat healthy

Instead of reaching for the chips and chocolate, fresh fruits and vegetables are not only yummy but offer nutritional benefits that help relieve stress. Some foods accelerate stress instead of reducing it and should be avoided. Foods that add to higher anxiety include artificial sweeteners, candy and other sugary foods, carbonated sodas, chips, chocolate, eggs, fried foods, junk foods, processed foods, red meat, and other foods containing preservatives or heavy spices. Instead, eat a bowl of berries and some yogurt!

Super Foods for Studying

https://www.test-preparation.ca/super-foods-studying/

Get plenty of ZZZZZZZs

Do not cram or try to do an all-nighter. If you created a study schedule at the beginning, and if you have stuck with that schedule, have confidence! Staying up too late trying to cram in last-minute bits of information is going to leave you exhausted the next day. Besides, whatever new information you cram in will only displace all the important ideas you've spent weeks learning. Remember: You need to be alert and fully functional the day of the exam

Have confidence in yourself!

Everyone experiences some anxiety when taking a test, but exhibiting a positive attitude banishes anxiety and fills you with the knowledge you really do know what you need to know. This is your opportunity to show how well prepared you are. Go for it!

Do not chitchat with friends

Let your friends know ahead of time that it is not anything personal, but you are going to ignore them in the test room! You need to find a seat away from doors and windows, one that has good lighting, and get comfortable. If other students are worried their anxiety could be detrimental to you; of course, you do not have to tell your friends that. If you are afraid they will be offended, tell them you are protecting them from your anxiety!

Common Test-Taking Mistakes

Taking a test is not much fun at best. When you take a test and make a stupid mistake that negatively affects your grade, it is natural to be very upset, especially when it is something that could have been easily avoided. So what are some of the common mistakes that are made on tests?

Do not fail to put your name on the test

How could you possibly forget to put your name on a test? You would be amazed at how often that happens. Very often, tests without names are thrown out immediately, resulting in a failing grade.

Marking the wrong multiple-choice answer

It is important to work at a steady pace, but that does not mean bolting through the questions. Be sure the answer you are marking is the one you mean to. If the bubble you need to fill in or the answer you need to circle is 'C', do not allow yourself to get distracted and select 'B' instead.

Answering a question twice

Some multiple-choice test questions have two very similar answers. If you are in too much of a hurry, you might select them both. Remember that only one answer is correct, so if you choose more than one, you have automatically failed that question.

Mishandling a difficult question

We recommend skipping difficult questions and returning to them later, but beware! First, be certain that you do return to the question. Circling the entire passage or placing a large question mark beside it will help you spot it when you are reviewing your test. Secondly, if you are not careful to skip the question, you can mess yourself up badly. Imagine that a question is too difficult and you decide to save it for later. You read the next question, which you know the answer, and you fill in that answer. You continue to the end of the test then return to the difficult question only to discover you didn't actually skip it! Instead, you inserted the answer to the following question in the spot reserved for the harder one, thus throwing off the remainder of your test!

Incorrectly Transferring an answer from scratch paper

This can happen easily if you are trying to hurry! Double check any answer you have figured out on scratch paper, and make sure what you have written on the test itself is an exact match!

Thinking too much

Oftentimes, your first thought is your best thought. If you worry yourself into insecurity, your self-doubts can trick you into choosing an incorrect answer when your first impulse was the right one!

Conclusion

CONGRATULATIONS! You have made it this far because you have applied yourself diligently to practicing for the exam and no doubt improved your potential score considerably! Passing your up-coming exam is a huge step in a journey that might be challenging at times but will be many times more rewarding and fulfilling. That is why being prepared is so important.

Good Luck!

Register for Free Updates and More Practice Test Questions

Register your purchase at

https://www.test-preparation.ca/register/ for fast and convenient access to updates, errata, free test tips and more practice test questions.

Online Resources

How to Prepare for a Test - The Ultimate Guide

https://www.test-preparation.ca/prepare-test/

Learning Styles - The Complete Guide

https://www.test-preparation.ca/learning-style/

Test Anxiety Secrets!

https://www.test-preparation.ca/test-anxiety/

Time Management on a Test

https://www.test-preparation.ca/time-management/

Flash Cards - The Complete Guide

https://www.test-preparation.ca/flash-cards/

Test Preparation Video Series

https://www.test-preparation.ca/test-video/

How to Memorize - The Complete Guide

https://www.test-preparation.ca/memorize/

www.ingramcontent.com/pod-product-compliance
Lightning Source LLC
LaVergne TN
LVHW052306200926
848481LV00002B/481